Lecture Notes in Applied and Computational Mechanics

Volume 50

Series Editors

Prof. Dr.-Ing. Friedrich Pfeiffer
Prof. Dr.-Ing. Peter Wriggers

Lecture Notes in Applied and Computational Mechanics

Edited by F. Pfeiffer and P. Wriggers

Further volumes of this series found on our homepage: springer.com

Vol.50: Ganghoffer, J.-F., Pastrone, F. (Eds.)
Mechanics of Microstructured Solids 2
102 p. 2010 [978-3-642-05170-8]

Vol. 49: Hazra, S.B.
Large-Scale PDE-Constrained Optimization in Applications
224 p. 2010 [978-3-642-01501-4]

Vol. 48: Su, Z.; Ye, L.
Identification of Damage Using Lamb Waves
346 p. 2009 [978-1-84882-783-7]

Vol. 47: Studer, C.
Numerics of Unilateral Contacts and Friction
191 p. 2009 [978-3-642-01099-6]

Vol. 46: Ganghoffer, J.-F., Pastrone, F. (Eds.)
Mechanics of Microstructured Solids
136 p. 2009 [978-3-642-00910-5]

Vol. 45: Shevchuk, I.V.
Convective Heat and Mass Transfer in Rotating Disk
Systems
300 p. 2009 [978-3-642-00717-0]

Vol. 44: Ibrahim R.A., Babitsky, V.I., Okuma, M. (Eds.)
Vibro-Impact Dynamics of Ocean Systems and Related
Problems
280 p. 2009 [978-3-642-00628-9]

Vol.43: Ibrahim, R.A.
Vibro-Impact Dynamics
312 p. 2009 [978-3-642-00274-8]

Vol. 42: Hashiguchi, K.
Elastoplasticity Theory
432 p. 2009 [978-3-642-00272-4]

Vol. 41: Browand, F., Ross, J., McCallen, R. (Eds.)
Aerodynamics of Heavy Vehicles II: Trucks, Buses,
and Trains
486 p. 2009 [978-3-540-85069-4]

Vol. 40: Pfeiffer, F.
Mechanical System Dynamics
578 p. 2008 [978-3-540-79435-6]

Vol. 39: Lucchesi, M., Padovani, C., Pasquinelli, G., Zani, N.
Masonry Constructions: Mechanical
Models and Numerical Applications
176 p. 2008 [978-3-540-79110-2]

Vol. 38: Marynowski, K.
Dynamics of the Axially Moving Orthotropic Web
140 p. 2008 [978-3-540-78988-8]

Vol. 37: Chaudhary, H., Saha, S.K.
Dynamics and Balancing of Multibody Systems
200 p. 2008 [978-3-540-78178-3]

Vol. 36: Leine, R.I.; van de Wouw, N.
Stability and Convergence of Mechanical Systems
with Unilateral Constraints
250 p. 2008 [978-3-540-76974-3]

Vol. 35: Acary, V.; Brogliato, B.
Numerical Methods for Nonsmooth Dynamical Systems:
Applications in Mechanics and Electronics
545 p. 2008 [978-3-540-75391-9]

Vol. 34: Flores, P.; Ambrósio, J.; Pimenta Claro, J.C.;
Lankarani Hamid M.
Kinematics and Dynamics of Multibody Systems
with Imperfect Joints: Models and Case Studies
186 p. 2008 [978-3-540-74359-0

Vol. 33: Nies ony, A.; Macha, E.
Spectral Method in Multiaxial Random Fatigue
146 p. 2007 [978-3-540-73822-0]

Vol. 32: Bardzokas, D.I.; Filshtinsky, M.L.;
Filshtinsky, L.A. (Eds.)
Mathematical Methods in Electro-Magneto-Elasticity
530 p. 2007 [978-3-540-71030-1]

Vol. 31: Lehmann, L. (Ed.)
Wave Propagation in Infinite Domains
186 p. 2007 [978-3-540-71108-7]

Vol. 30: Stupkiewicz, S. (Ed.)
Micromechanics of Contact and Interphase Layers
206 p. 2006 [978-3-540-49716-5]

Vol. 29: Schanz, M.; Steinbach, O. (Eds.)
Boundary Element Analysis
571 p. 2006 [978-3-540-47465-4]

Vol. 28: Helmig, R.; Mielke, A.; Wohlmuth, B.I. (Eds.)
Multifield Problems in Solid and Fluid Mechanics
571 p. 2006 [978-3-540-34959-4

Vol. 27: Wriggers P., Nackenhorst U. (Eds.)
Analysis and Simulation of Contact Problems
395 p. 2006 [978-3-540-31760-9]

Mechanics of Microstructured Solids 2

Cellular Materials, Fibre Reinforced Solids and Soft Tissues

Jean-François Ganghoffer and
Franco Pastrone (Eds.)

 Springer

Prof. Jean-François Ganghoffer
LEMTA - ENSEM, France
2, Avenue de la Forêt de Haye
BP 160 - 54504 Vandoeuvre Cedex
France
E-mail: Jean-francois.Ganghoffer@ensem.inpl-nancy.fr

Prof. Franco Pastrone
Department of Mathematics
University of Torino
Via Carlo Alberto, 10
10123 Torino
Italy
E-mail: franco.pastrone@unito.it

ISBN: 978-3-642-26227-2   e-ISBN: 978-3-642-05171-5

DOI 10.1007/ 978-3-642-05171-5

Lecture Notes in Applied and Computational Mechanics   ISSN 1613-7736

e-ISSN 1860-0816

© Springer-Verlag Berlin Heidelberg 2010
Softcover reprint of the hardcover 1st edition 2010

This work is subject to copyright. All rights are reserved, whether the whole or part of the material is concerned, specifically the rights of translation, reprinting, reuse of illustrations, recitation, broadcasting, reproduction on microfilm or in any other ways, and storage in data banks. Duplication of this publication or parts thereof is permitted only under the provisions of the German Copyright Law of September 9, 1965, in its current version, and permission for use must always be obtained from Springer. Violations are liable for prosecution under the German Copyright Law.

The use of general descriptive names, registered names, trademarks, etc. in this publication does not imply, even in the absence of a specific statement, that such names are exempt from the relevant protective laws and regulations and therefore free for general use.

Typeset & Cover Design: Scientific Publishing Services Pvt. Ltd., Chennai, India.

Printed on acid-free paper

9 8 7 6 5 4 3 2 1 0

springer.com

Forword

This second volume of the series *Lecture Notes in Applied and Computational Mechanics* is the second part of the compendium of reviewed articles presented at the 11th EUROMECH-MECAMAT conference entitled "Mechanics of microstructured solids: cellular materials, fibre reinforced solids and soft tissues", which took place in Torino (Italy) in March 10–14, 2008, at the Museo Regional delle Scienze. This EUROMECH-MECAMAT conference was jointly organized by the Dipartimento di Matematica dell'Università di Torino, Italy and the INPL Institute (LEMTA, Nancy-Université, France). Prof. Franco Pastrone and Prof. Jean-François Ganghoffer were the co-chairmen.

The conference brought together 50 scientists from 11 European countries, and was aimed at defining the current state of the art in the growing field of cellular and fibrous materials in Europe. Participants had interests in the constitutive models of micro-structured solids, non-linear wave propagation, setting up of models and identification of fibre reinforced solids, and soft tissue behaviour in a biomechanical context.

The conference covered most of the mechanical and material aspects, grouped in the following four sessions:

- Fibre reinforced materials;
- Soft biological tissues;
- Generalized continua: models and materials;
- Non-linear wave propagation.

The high quality talks showed a good balance between modelling and material aspects. An important part of the colloquium, with 12 presentations, was devoted to various aspects of the biomechanics of soft tissues, such as cell adhesion, constitutive models of soft tissues (brain; arteries), or models of blood flow.

Beyond the scientific reports and the poster presentations, many informal exchanges were possible mainly during the coffee breaks and the lunches, according to the scope and the spirit of Euromech Colloquia and Conferences. Surely the meeting stimulated current researches and often the discussions were lively, informal and penetrating.

Contents

Linear Elasticity with Couple Stresses 1
Manfred Braun

Vectorial Microstructures: Applications to Fabrics and
Granular Media 9
Franco Pastrone

Numerical Simulation of Interaction of Solitons and Solitary
Waves in Granular Materials 21
Andrus Salupere, Lauri Ilison

Multi-scale Modelling of Fracture in Open-Cell Metal
Foams 29
K.R. Mangipudi, P.R. Onck

AFCs: Active-Stress vs. Active-Strain Modeling 37
Paola Nardinocchi, Paolo Podio-Guidugli

On Eshelby Tensors, Thermodynamics and Calculus of
Variations 49
Jean-François Ganghoffer

Nonlinear Hyperbolic Equations and Linear Heat Conduction
with Memory 63
Sandra Carillo

Mesoscopic Mechanical Analyses of Textile Composites:
Validation with X-Ray Tomography 71
Pierre Badel, Eric Maire, Emmanuelle Vidal-Sallé, Philippe Boisse

Mechanical Response of Helically Wound Fiber-Reinforced
Incompressible Non-linearly Elastic Pipes 79
Paola Nardinocchi, Tomas Svaton, Luciano Teresi

Author Index 89

Linear Elasticity with Couple Stresses

Manfred Braun

Summary. The linear couple-stress theory of Mindlin and Tiersten is reformulated allowing also for dislocations and disclinations. The governing equations are compiled in form of a Tonti diagram.

1 Introduction

The linear theory of elasticity is based on the assumption that only the symmetric strain tensor is responsible for strain energy and stress in the body. The skew-symmetric part of the displacement gradient represents a rigid-body rotation of the material elements and, according to the principle of material frame indifference, must not occur as an argument of the constitutive equations. The simplest generalization of ordinary elasticity theory is to include the gradient of the local rotation as a source of stress and strain energy, see [3]. This means that strain energy may depend not only on first-order derivatives of the displacement field but also on *some* derivatives of second order. In this sense, the theory of Mindlin and Tiersten can be interpreted as a "higher-order gradient theory" in which, however, only the gradient of the rotation but not the gradient of strain is considered.

The aim of this paper is to reformulate Mindlin and Tiersten's theory, and to provide a systematic compilation of the governing equations. Following Tonti [4] the equations are arranged in a diagram which points out their interrelations and reveals the duality between geometric and stress equations.

2 Geometry of Deformation

The theory starts from the displacement field $u(x)$ within the body. Its gradient

$$U = \nabla u \tag{1}$$

Manfred Braun
Universität Duisburg-Essen, Lehrstuhl für Mechanik und Robotik,
47048 Duisburg, Germany
e-mail: `manfred.braun@uni-due.de`

J.-F. Ganghoffer, F. Pastrone (Eds.): Mech. of Microstru. Solids 2, LNACM 50, pp. 1–8.
springerlink.com © Springer-Verlag Berlin Heidelberg 2010

is decomposed into its symmetric and skew-symmetric parts,

$$U = E + \omega \times I, \tag{2}$$

where

$$E = \operatorname{sym} U \quad \text{and} \quad \omega = \frac{1}{2} I \times U \tag{3}$$

denote the symmetric strain tensor and the rotation vector, respectively. The outer products $\omega \times I$ and $I \times U$ are understood in the sense of de Boer [2].

In classical elasticity theory, strain energy and stress depend on the strain tensor E only and must not depend on the rotation vector ω, as this would contradict the principle of frame indifference. However, the gradient of the rotation vector may occur as an argument in the constitutive equations. Therefore the *structural curvature tensor*

$$K = \nabla \omega \tag{4}$$

is introduced. Strain energy, which now should be called "strain-and-curvature energy", is assumed to depend on both the symmetric strain tensor E and the structural curvature tensor K which, in general, has no inherent symmetry property. Since $\omega = \frac{1}{2} \operatorname{curl} u$, the curvature tensor is deviatoric, i. e., $\operatorname{tr} K = 0$.

3 Constitutive Equations

In the classical linear theory of elasticity, the strain-energy density is a quadratic form in the strain components. If the structural curvature tensor (4) is considered as an additional argument, the strain energy is provided by the quadratic form

$$W = \frac{1}{2} \left(A_{ijkl} E_{ij} E_{kl} + 2 B_{ijkl} E_{ij} K_{kl} + C_{ijkl} K_{ij} K_{kl} \right). \tag{5}$$

The coefficients A_{ijkl}, B_{ijkl}, and C_{ijkl} have different physical dimensions and satisfy certain symmetry relations. Since E is symmetric and K is deviatoric there are 105 independent elastic constants, in general. The 21 constants of standard anisotropic elasticity, A_{ijkl}, are complemented by 48 constants B_{ijkl} and 36 constants C_{ijkl}.

In the isotropic case, the elasticity tensors are expressible in terms of 5 independent material parameters, in general. Another assumption about material behavior is *centrosymmetry*, which means that the strain-energy density is invariant under a central inversion $x \mapsto -x$ of the material points. Under this transformation the strain tensor keeps its sign, $E \mapsto E$, while the structural curvature tensor reverts it, $K \mapsto -K$. Invariance of the strain energy under central inversion requires $B_{ijkl} = 0$. If the material is both isotropic and centrosymmetric as will be assumed in the sequel, its strain-energy density has the form

$$W = \frac{1}{2} \left[\lambda \left(\operatorname{tr} E \right)^2 + 2\mu E \cdot E + \bar{\bar{\mu}}_1 K \cdot K + \bar{\bar{\mu}}_2 K \cdot K^\mathsf{T} \right] \tag{6}$$

with only 4 relevant elastic parameters.

4 Equilibrium Conditions

The total strain energy of a body $\mathcal{B}$ is

$$\Pi = \int_{\mathcal{B}} W(\boldsymbol{E}, \boldsymbol{K})\, \mathrm{d}V. \tag{7}$$

In a state of equilibrium, it attains a stationary value with respect to variations of the strain and curvature fields, $\boldsymbol{E}(\boldsymbol{x})$ and $\boldsymbol{K}(\boldsymbol{x})$, satisfying the geometric equations (1), (3) and (4), along with suitable boundary conditions. The restrictions can be incorporated into the strain-energy functional by use of Lagrange multipliers. Thus instead of (7), the augmented strain-energy

$$\Pi^* = \int_{\mathcal{B}} \Big[W(\boldsymbol{E}, \boldsymbol{K}) + \boldsymbol{T} \cdot (\operatorname{grad} \boldsymbol{u} - \boldsymbol{U}) + \boldsymbol{S} \cdot (\operatorname{sym} \boldsymbol{U} - \boldsymbol{E}) + $$
$$+ \boldsymbol{t} \cdot (\operatorname{ax} \boldsymbol{U} - \boldsymbol{\omega}) + \boldsymbol{M} \cdot (\operatorname{grad} \boldsymbol{\omega} - \boldsymbol{K}) \Big]\, \mathrm{d}V \tag{8}$$

is introduced, with the symmetric tensor field $\boldsymbol{S}$, the tensor field $\boldsymbol{M}$, and the axial vector field $\boldsymbol{t}$ as Lagrange multipliers.

Using partial integration the variation of the augmented strain energy is obtained as

$$\delta \Pi^* = \int_{\partial \mathcal{B}} (\delta \boldsymbol{u} \cdot \boldsymbol{T} + \delta \boldsymbol{\omega} \cdot \boldsymbol{M})\, \mathrm{d}a + $$
$$+ \int_{\mathcal{B}} \Big[\Big(\boldsymbol{S} + \tfrac{1}{2} \boldsymbol{t} \times \boldsymbol{I} - \boldsymbol{T} \Big) \cdot \delta \boldsymbol{U} - (\operatorname{div} \boldsymbol{M} - \boldsymbol{t}) \cdot \delta \boldsymbol{\omega} - (\operatorname{div} \boldsymbol{T}) \cdot \delta \boldsymbol{u} + $$
$$+ \Big(\frac{\partial W}{\partial \boldsymbol{E}} - \boldsymbol{S} \Big) \cdot \delta \boldsymbol{E} + \Big(\frac{\partial W}{\partial \boldsymbol{K}} - \boldsymbol{M} \Big) \cdot \delta \boldsymbol{K} \Big]\, \mathrm{d}V. \tag{9}$$

If no volume force is impressed within the body the volume integral has to vanish identically for arbitrary variations $\delta \boldsymbol{E}$, $\delta \boldsymbol{K}$, $\delta \boldsymbol{U}$, $\delta \boldsymbol{u}$ and $\delta \boldsymbol{\omega}$.

The variations $\delta \boldsymbol{E}$ and $\delta \boldsymbol{K}$ yield the constitutive equations

$$\boldsymbol{S} = \frac{\partial W}{\partial \boldsymbol{E}} \quad \text{and} \quad \boldsymbol{M} = \frac{\partial W}{\partial \boldsymbol{K}}, \tag{10}$$

which identify the Lagrange multipliers $\boldsymbol{S}$ and $\boldsymbol{M}$ as the symmetric stress and the couple stress tensors, respectively. From the variation $\delta \boldsymbol{U}$ the decomposition of total stress

$$\boldsymbol{T} = \boldsymbol{S} + \frac{1}{2} \boldsymbol{t} \times \boldsymbol{I} \tag{11}$$

into its symmetric and skew-symmetric parts is obtained. The variations $\delta \boldsymbol{u}$ and $\delta \boldsymbol{\omega}$ result in the equilibrium conditions

$$\operatorname{div} \boldsymbol{T} = \boldsymbol{0} \quad \text{and} \quad \operatorname{div} \boldsymbol{M} + \boldsymbol{t} = \boldsymbol{0} \tag{12}$$

of stress and couple stress, respectively.

The variation of strain energy under the geometric restrictions (3) and (4) has brought forward the constitutive equations (10), the decomposition of stress (11), and the equilibrium conditions (12) thus completing the set of governing equations.

5 Navier's Equation and Kelvin's Problem

Navier's equation is the partial differential equation that governs the displacement field $\boldsymbol{u}$. In classical elasticity theory it is a partial differential equation of second order. By taking into account couple stresses the order is enhanced to four.

The constitutive equations for stress and couple stress in an isotropic, centrosymmetric material are obtained by applying (10) to the strain-energy density (6) which yields

$$\boldsymbol{S} = \lambda(\operatorname{tr}\boldsymbol{E})\boldsymbol{I} + 2\mu\boldsymbol{E} \quad\text{and}\quad \boldsymbol{M} = \bar{\bar{\mu}}_1\boldsymbol{K} + \bar{\bar{\mu}}_2\boldsymbol{K}^{\mathsf{T}}. \tag{13}$$

Inserting the definitions of strain and curvature one further obtains

$$\boldsymbol{S} = \lambda(\operatorname{div}\boldsymbol{u})\boldsymbol{I} + \mu\nabla\boldsymbol{u} + \mu\left(\nabla\boldsymbol{u}\right)^{\mathsf{T}},$$
$$\boldsymbol{M} = \frac{1}{2}\bar{\bar{\mu}}_1\nabla\operatorname{curl}\boldsymbol{u} + \frac{1}{2}\bar{\bar{\mu}}_2\left(\nabla\operatorname{curl}\boldsymbol{u}\right)^{\mathsf{T}}. \tag{14}$$

The equilibrium of forces, $(12)_1$, pertains to the total stress (11) composed of the symmetric stress $\boldsymbol{S}$ and the antisymmetric stress $\frac{1}{2}\boldsymbol{t}\times\boldsymbol{I}$. Recalling the equilibrium of couple stress, $(12)_2$, yields

$$\operatorname{div}(\boldsymbol{t}\times\boldsymbol{I}) = -\operatorname{curl}\boldsymbol{t} = \operatorname{curl}\operatorname{div}\boldsymbol{M}. \tag{15}$$

By inserting the explicit representation of the stress and couple stress tensors, (14), one finally obtains Navier's equation in the form

$$(\lambda+\mu)\nabla\operatorname{div}\boldsymbol{u} + \mu\Delta\boldsymbol{u} + \frac{1}{4}\bar{\bar{\mu}}_1\Delta\left(\nabla\operatorname{div}\boldsymbol{u} - \Delta\boldsymbol{u}\right) = 0. \tag{16}$$

The first two terms correspond to the classical Navier's equation. The generalized version contains only one additional material constant $\bar{\bar{\mu}}_1$. The number of independent parameters in (16) can still be reduced. Division by the shear modulus μ yields

$$\frac{1}{1-2\nu}\nabla\operatorname{div}\boldsymbol{u} + \Delta\boldsymbol{u} + \ell^2\Delta\left(\nabla\operatorname{div}\boldsymbol{u} - \Delta\boldsymbol{u}\right) = 0, \tag{17}$$

with the new material parameters

$$\nu = \frac{\lambda}{2(\lambda+\mu)} \quad\text{and}\quad \ell^2 = \frac{\bar{\bar{\mu}}_2}{\mu}. \tag{18}$$

In addition to Poisson's ratio ν a length scale ℓ occurs in the final form (17) of Navier's equation.

Kelvin's problem consists in finding the displacement field in an elastic space loaded by a concentrated force acting at the origin. Thus the displacement field $\boldsymbol{u}(\boldsymbol{x})$ has to satisfy Navier's equation (17) everywhere except at $\boldsymbol{x} = \boldsymbol{0}$. Due to the linearity of the underlying theory, the displacement field can be assumed in the form

$$\boldsymbol{u} = \boldsymbol{G}(\boldsymbol{x})\boldsymbol{P}, \tag{19}$$

where $\boldsymbol{P}$ denotes the force vector applied at the origin. The Green's tensor $\boldsymbol{G}$ must be an isotropic tensor function of the position vector $\boldsymbol{x}$. Its most general form is

$$\boldsymbol{G}(\boldsymbol{x}) = \frac{1}{4\pi\mu} \left[\varphi(x)\boldsymbol{I} + \psi(x)\boldsymbol{e} \otimes \boldsymbol{e} + \chi(x)\boldsymbol{e} \times \boldsymbol{I} \right], \tag{20}$$

where

$$x = |\boldsymbol{x}| \quad \text{and} \quad \boldsymbol{e} = \frac{1}{x}\boldsymbol{x} \tag{21}$$

denote the length of the position vector and the unit vector aligned with it. If the material is centrosymmetric the last term in (20) vanishes, $\chi(x) \equiv 0$, because the force $\boldsymbol{P}$ cannot generate an axial rotation.

The functions $\varphi(x)$ and $\psi(x)$ must be of physical dimension 1/Length. In detail they have to be determined from Navier's equation (17) and integrated boundary conditions at a sphere surrounding the origin. A lengthy but straightforward calculation yields the functions

$$\varphi(x) = \frac{1}{x} \left[1 - c + \frac{\ell^2}{x^2} - \left(1 + \frac{\ell}{x} + \frac{\ell^2}{x^2} \right) \exp\left(-\frac{x}{\ell} \right) \right]$$

$$\psi(x) = \frac{1}{x} \left[c - \frac{3\ell^2}{x^2} + \left(1 + \frac{3\ell}{x} + \frac{3\ell^2}{x^2} \right) \exp\left(-\frac{x}{\ell} \right) \right], \tag{22}$$

where

$$c = \frac{1}{4(1 - \nu)} \tag{23}$$

denotes a dimensionless constant. For vanishing scale factor ℓ, the functions are reduced to the well known result of classical elasticity.

6   Compatibility Conditions

The displacement field $\boldsymbol{u}(\boldsymbol{x})$ is obtained by integrating the displacement gradient $\boldsymbol{U} = \boldsymbol{E} + \boldsymbol{\omega} \times \boldsymbol{I}$. The integral has to be path independent, which means that

$$\int_C (\boldsymbol{E} + \boldsymbol{\omega} \times \boldsymbol{I}) \, \mathrm{d}\boldsymbol{x} = \boldsymbol{0} \tag{24}$$

has to hold for any *closed* path $\mathcal{C} = \partial\mathcal{A}$. Using Stokes' theorem

$$\int_{\partial\mathcal{A}} (\boldsymbol{E} + \boldsymbol{\omega} \times \boldsymbol{I})\, \mathrm{d}\boldsymbol{x} = \int_{\mathcal{A}} [\operatorname{curl}\boldsymbol{E} + \operatorname{curl}(\boldsymbol{\omega} \times \boldsymbol{I})]^{\mathsf{T}}\, \mathrm{d}\boldsymbol{a} \qquad (25)$$

the path integral is transformed into a surface integral whose integrand has to vanish identically. Using the identity

$$\operatorname{curl}(\boldsymbol{\omega} \times \boldsymbol{I}) = (\operatorname{div}\boldsymbol{\omega})\boldsymbol{I} - \operatorname{grad}\boldsymbol{\omega} = (\operatorname{tr}\boldsymbol{K})\boldsymbol{I} - \boldsymbol{K}, \qquad (26)$$

one arrives at the first compatibility condition

$$\boldsymbol{K} = \operatorname{curl}\boldsymbol{E} - \frac{1}{2}\operatorname{tr}(\operatorname{curl}\boldsymbol{E}) = \operatorname{curl}\boldsymbol{E}. \qquad (27)$$

The last equality is due to the fact that $\operatorname{tr}(\operatorname{curl}\boldsymbol{E}) \equiv 0$.

In a similar way, the rotation vector is obtained by integrating the structural curvature $\boldsymbol{K}$. To make the integration path independent the condition

$$\int_{\mathcal{C}} \boldsymbol{K}\, \mathrm{d}\boldsymbol{x} = \boldsymbol{0} \qquad (28)$$

has to hold for any closed path $\mathcal{C} = \partial\mathcal{A}$. Applying again Stokes' theorem

$$\int_{\partial\mathcal{A}} \boldsymbol{K}\, \mathrm{d}\boldsymbol{x} = \int_{\mathcal{A}} (\operatorname{curl}\boldsymbol{K})^{\mathsf{T}}\, \mathrm{d}\boldsymbol{a} \qquad (29)$$

one obtains the second compatibility condition

$$\operatorname{curl}\boldsymbol{K} = \boldsymbol{O}. \qquad (30)$$

It ensures that there is no closing error of rotation, while the first compatibility condition (27) ensures that there is no closing error of displacement.

The compatibility conditions (27) and (30) can be combined into one compatibility condition

$$\operatorname{inc}\boldsymbol{E} = \operatorname{curl}\operatorname{curl}\boldsymbol{E} = \boldsymbol{O}, \qquad (31)$$

where inc denotes Kröner's incompatibility operator. The compatibility condition has the same form as in classical elasticity. The geometry of deformation does not depend on which of the fields may serve as an argument of the strain-energy function.

Further identities can be obtained by allowing closing errors of displacement and rotation. The dislocation and disclination densities,

$$\boldsymbol{A} = (\operatorname{curl}\boldsymbol{E})^{\mathsf{T}} + \boldsymbol{I} \times\!\!\times \boldsymbol{K} \quad \text{and} \quad \boldsymbol{\Theta} = (\operatorname{curl}\boldsymbol{K})^{\mathsf{T}}, \qquad (32)$$

have to satisfy the compatibility conditions

$$\operatorname{div}\boldsymbol{A} + \boldsymbol{I} \times \boldsymbol{\Theta} = \boldsymbol{0} \quad \text{and} \quad \operatorname{div}\boldsymbol{\Theta} = \boldsymbol{0}, \qquad (33)$$

see [1]. These conclude the equations describing the geometry of deformation.

7 Stress Functions

The representation of the stress and couple stress tensors by stress functions is intimately connected with the compatibility conditions. In a state of equilibrium, the strain energy (7) attains a stationary value with respect to variations of $\boldsymbol{E}(\boldsymbol{x})$ and $\boldsymbol{K}(\boldsymbol{x})$ satisfying the compatibility conditions as constraints. If dislocations and disclinations are permitted, the variations of the fields $\boldsymbol{E}$, $\boldsymbol{K}$, $\boldsymbol{A}$ and $\boldsymbol{\Theta}$ are restricted by the equations (32) and (33). In the same way as in (8) an augmented strain energy

$$\Pi^{**} = \int_{\mathcal{B}} \left\{ W(\boldsymbol{E}, \boldsymbol{K}) + \boldsymbol{\Psi} \cdot \left[\boldsymbol{A} - (\operatorname{curl} \operatorname{sym} \boldsymbol{E})^{\mathsf{T}} - \boldsymbol{I} \times\!\!\!\times \boldsymbol{K} \right] + \right.$$
$$\left. + \boldsymbol{\Phi} \cdot \left[\boldsymbol{\Theta} - (\operatorname{curl} \boldsymbol{K})^{\mathsf{T}} \right] + \psi \cdot (\operatorname{div} \boldsymbol{A} + \boldsymbol{I} \times \boldsymbol{\Theta}) + \varphi \cdot \operatorname{div} \boldsymbol{\Theta} \right\} \, \mathrm{d}V \quad (34)$$

is formed, where the restrictions are incorporated by using Lagrange multipliers $\boldsymbol{\Psi}$, $\boldsymbol{\Phi}$, ψ and φ. From the variation of the augmented strain energy (34), after applying partial integration, the governing equations of the Lagrange multipliers can be read off. Thus the stress and couple stress tensors are represented by

$$\boldsymbol{S} = \operatorname{sym}(\operatorname{curl} \boldsymbol{\Psi})^{\mathsf{T}} \quad \text{and} \quad \boldsymbol{M} = (\operatorname{curl} \boldsymbol{\Phi})^{\mathsf{T}} + \boldsymbol{I} \times\!\!\!\times \boldsymbol{\Psi} \qquad (35)$$

in terms of stress function tensors $\boldsymbol{\Psi}$ and $\boldsymbol{\Phi}$. A stress-free state is generated from the stress function tensors

$$\boldsymbol{\Phi} = \operatorname{grad} \varphi + \boldsymbol{I} \times \psi \quad \text{and} \quad \boldsymbol{\Psi} = \operatorname{grad} \psi. \qquad (36)$$

These representations are the duals to the identities (33) satisfied by the dislocation and disclination densities.

8 Tonti Diagram

The governing equations of linear elasticity with couple stresses can be subdivided into (i) the geometric equations presented in Sections 2 and 6, and (ii) the stress equations presented in Sections 4 and 7. These two threads are connected by the constitutive equations of Section 3. Compiling all equations systematically leads in a natural way to the Tonti diagram as depicted in Figure 1.

Not all of the equations in the Tonti diagram have the same relevance. Since a state of zero stress is not of any interest one should omit the representation of the stress function tensors $\boldsymbol{\Phi}$, $\boldsymbol{\Psi}$ by the vector potentials φ, ψ on the lower right of the Tonti diagram. (The stress functions themselves may be regarded as purely auxiliary quantities without any descriptive mechanical interpretation.)

If one allows for a continuous distribution of dislocations and disclinations the notions of displacement and rotation lose their meaning, at least they are not uniquely defined as vector fields. In this case, the upper left part

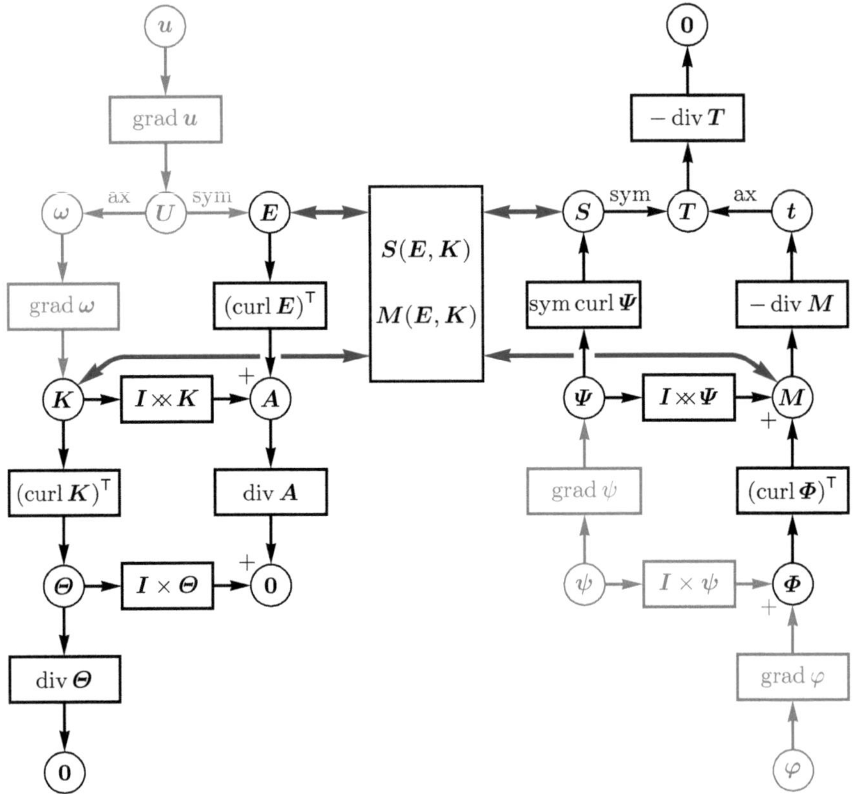

Fig. 1 Tonti diagram

of the Tonti diagram has to be omitted such that the geometry starts with the strain tensor E and the structural curvature tensor K as the independent quantities. In Figure 1 these "irrelevant" quantities and equations are depicted in gray.

References

1. Anthony, K.H.: Die Theorie der Disklinationen. Archive for Rational Mechanics and Analysis 39, 43–88 (1970)
2. de Boer, R.: Vektor- und Tensorrechnung für Ingenieure. Springer, Berlin (1982)
3. Mindlin, R.D., Tiersten, H.F.: Effects of couple-stresses in linear elasticity. Archive for Rational Mechanics and Analysis 11, 415–448 (1962)
4. Tonti, E.: On the mathematical structure of a large class of physical theories. Atti della Accademia Nazionale dei Lincei, Serie Ottava, Rendiconti, Classe di Scienze fisiche, matematiche e naturali 52, 48–56 (1972)

Vectorial Microstructures: Applications to Fabrics and Granular Media

Franco Pastrone

Abstract. Textile fabrics and 2D granular media can be modelled by means of non-classical nonlinear elasticity, in particular using the theory of vectorial microstructures. The field equations are obtained via a variational principle and suitable constitutive relations are given in both cases. The microstructure takes into account the micro-ondulation of the fibres in textiles and the grains in granular media.

Keywords: Complex materials, nonlinear models, microstructures, textiles, plane granular media.

1 Introduction

In this presentation two examples are provided of possible applications of the theory of nonlinear complex materials, within the framework of non-classical nonlinear elasticity, namely textiles and to 2D granular media.

The model of complex material is restricted here to the so called vectorial microstructures, which can encompass many relevant models, like Cosserat media, granular solids, micromorphic bodies, solids with micro cracks, textiles. Some interesting results have been obtained in the 1 dimensional case, when a scalar function describes the microstructure and both dispersion and dissipation can be taken in account, within, obviously, the nonlinearity of the constitutive equations, such that the field equations result in non linear PDE's. A wide class of phenomena can be described by means of microstructural models of solids and fluids, where the microstructure can be described by vector fields over the body. In principle, there are no restrictions on the number of vector fields, which are unknown variables of the problem, but

Franco Pastrone
Department of Mathematics, University of Torino,
Via Carlo Alberto 10, 10123 Torino, Italy
e-mail: `franco.pastrone@unito.it`

J.-F. Ganghoffer, F. Pastrone (Eds.): Mech. of Microstru. Solids 2, LNACM 50, pp. 9–20.
springerlink.com
© Springer-Verlag Berlin Heidelberg 2010

there are obvious restrictions due to the possible physical meaning of each vector field. In the Section 2 the field equations and the constitutive relations of an elastic solid with microstructure are obtained. In the Section 3 is introduced a suitable model of a textile fabric, in which the vectorial microstructure is related to the microondulation of the weft and the warp. We assume that the fabric is a surface with bending stiffness and in which the effects due to the microundulations of the welft and the warp are taken into account. In the Section 4 we apply the model with microstructures to plane granular solids, and again non linear field equations are derived.

2 The Field Equations

Let B be the body, considered as a manifold embedded in a 3-dimensional affine space, $\mathbf{X}$ a point of this body in its reference configuration C^*, and $\mathbf{x}$ the corresponding point in the actual configuration C. As usual, the displacement is given, in terms of the position vector $\mathbf{r} = \mathbf{r}(\mathbf{X}^h, t) = \mathbf{x} - \mathbf{0}$, $\mathbf{0}$ a fixed point in the physical space and $\mathbf{R} = \mathbf{X} - \mathbf{0}$, by $\mathbf{u} = \mathbf{r} - \mathbf{R}$. Commas denote partial derivatives with respect to X^h and superposed dots denote partial derivatives with respect to time, e.g.:

$$\mathbf{r}_{,h} = \frac{\partial \mathbf{u}}{\partial X^h}; \quad \dot{\mathbf{r}} = \frac{\partial \mathbf{u}}{\partial t}.$$

By microstructure, we mean that at each point $x \in C$ it is possible to apply a microscope and discover a "small world". We assume that this "small world" is a manifold of dimension n, and we label this micro-manifold by M_x. By means of a mathematical procedure, elsewhere called "magnification process" (see Pastrone [10, 11, 13]) which generalises the linear model by Mindlin (1964), one can associate to this microscopic world a set of vectors,

$$\mathbf{d}_K = \mathbf{d}_K(X^h, t) \quad K = 1, 2, ..., N$$

Such vectors can be called "directors", according to the usual language of polar continua, and they are apt to provide a description of some properties of the microstructure as they act at the macroscopic level.

The kinetic energy of a microstructured body is defined as a quadratic form in the velocities $\dot{\mathbf{r}}$, $\dot{\mathbf{d}}_H$:

$$T = \frac{1}{2}\left(\rho\dot{\mathbf{r}} \cdot \dot{\mathbf{r}} + 2\rho^H \dot{\mathbf{r}} \cdot \dot{\mathbf{d}}_H + \rho^{HK}\dot{\mathbf{d}}_H \cdot \dot{\mathbf{d}}_K\right) \tag{2.1}$$

where ρ, ρ^H, ρ^{HK} are functions of $\mathbf{x}$ and t:

T is a definite positive quadratic form: $T \geq 0$, $T = 0 \Leftrightarrow \dot{\mathbf{r}} = \dot{\mathbf{d}} = 0$.

Hence, without any loss of generality, we can reduce it to a diagonal form:

$$T = \frac{1}{2}\rho\left(\dot{\mathbf{r}} \cdot \dot{\mathbf{r}} + I^{HK}\dot{\mathbf{d}}_H \cdot \dot{\mathbf{d}}_K\right) \tag{2.2}$$

where ρ is the mass density and I^{HK} represents the inertia form of the microstructure.

If we deal with a Lagrangian formulation, ρ and I are defined on the reference configuration, hence they are functions of $\mathbf{X}$ only.

Let us assume that the body admits a generalized stored energy density:

$$W = W(\mathbf{r}_{,h}, \mathbf{d}_H, \mathbf{d}_{H,J}, \mathbf{X}) \tag{2.3}$$

which is related to the total mechanical power expended of any motion through the equation:

$$P_T = \frac{dW}{dt} = \frac{\partial W}{\partial \mathbf{r}_{,h}}\dot{\mathbf{r}}_{,h} + \frac{\partial W}{\partial \mathbf{d}_H}\dot{\mathbf{d}}_H + \frac{\partial W}{\partial \mathbf{d}_{H,i}}\dot{bd}_{H,i}. \tag{2.4}$$

where $\dot{\mathbf{r}}_{,h}$; $\dot{\mathbf{d}}_H$; $\dot{\mathbf{d}}_{H,J}$ are the strain velocities and $\dfrac{\partial W}{\partial \mathbf{r}_{,h}}$; $\dfrac{\partial W}{\partial \mathbf{d}_H}$; $\dfrac{\partial W}{\partial \mathbf{d}_{H,J}}$ represents the generalized stresses.

We will also take in account conservative body forces such that there exists a potential

$$W_b = W_b(\mathbf{r}, \mathbf{d}_H, \mathbf{X})$$

hence the power of the body forces is given by:

$$P_T^b = \frac{\partial W_b}{\partial \mathbf{r}} \cdot \dot{\mathbf{r}} + \frac{\partial W_b}{\partial \mathbf{d}_H} \cdot \dot{\mathbf{d}}_H$$

We can drive the fields equation via an usual variational principle, namely requiring that the motion of the body in a certain interval of time $[t_0, t_1]$ will make to vanish the energy fnctional, among all admissible motion:

$$\mathcal{E} = \int_{t_0}^{t_1} \left[\int_{\mathcal{B}} (T - W - W_b) dV \right] dt. \tag{2.5}$$

The corresponding Euler-Lagrange equations read:

$$\begin{cases} \left(\dfrac{\partial W}{\partial \boldsymbol{r}_{,i}} \right)_{,i} - \dfrac{\partial W_b}{\partial \boldsymbol{r}} = \dfrac{d}{dt} \dfrac{\partial T}{\partial \dot{\boldsymbol{r}}} \\[4mm] \left(\dfrac{\partial W}{\partial \boldsymbol{d}_{H,i}} \right)_{,i} - \dfrac{\partial W}{\partial \boldsymbol{d}_H} - \dfrac{\partial W_b}{\partial \boldsymbol{d}_H} = \dfrac{d}{dt} \dfrac{\partial T}{\partial \dot{\boldsymbol{d}}_H} \end{cases} \tag{2.6}$$

In general, physically, the microbody forces are different from macrobody forces, hence we can split W_b in two parts:

$$W_b = W_b^{macro}(\boldsymbol{r}, \underline{\mathbf{X}}) + W_b^{micro}(\boldsymbol{d}_H, \mathbf{X}) \tag{2.7}$$

In (2.7) $W_b^{macro}(\boldsymbol{r}, \mathbf{X})$ represents the potential of the macrobody forces, while $W_b^{micro}(\boldsymbol{d}_H, \mathbf{X})$ represents the potential of the microbody forces, since we assume that both forces be conservative. Let us remark that the macroforces depend on the position vector and the microforces on the microvector fields. Both can depend on the material coordinates.

From a formal point of view, we can recover the field equations in a classical form (Euler equations) introducing a set of forces, coupling moments, generalized forces and moments as follows:

$$\begin{cases} \boldsymbol{\sigma}^i = \dfrac{\partial W}{\partial \boldsymbol{r}_{,i}} \; ; \; \boldsymbol{\eta}^{Hi} = \dfrac{\partial W}{\partial \boldsymbol{d}_{H,i}} \; ; \; \boldsymbol{\tau}^H = -\dfrac{\partial W}{\partial \boldsymbol{d}_H} \\[3mm] \boldsymbol{B} = -\dfrac{\partial W_b^{macro}}{\partial \boldsymbol{r}} \; ; \; \boldsymbol{b}_H^{micro} = -\dfrac{\partial W_b^{micro}}{\partial \boldsymbol{d}_H} \end{cases} \tag{2.8}$$

The balance of moment of momentum, which in the classical theory implies that the Cauchy stress is a symmetric tensor, leads in our model to the algebraic equations:

$$\mathbf{r}_{,i} \times \boldsymbol{\sigma}^i - \mathbf{d}_H \times \boldsymbol{\tau}^H + \mathbf{d}_{H,j} \times \boldsymbol{\eta}^{Hj} = \mathbf{0} \tag{2.9}$$

In order to obtain (2.9), as proved in details in [10], we make use of the objectivity of W:

$$W(\mathbf{r}_{,i}; \mathbf{d}_H; \mathbf{d}_{H,j}; \mathbf{X}) = W(Q\mathbf{r}_{,i}; Q\mathbf{d}_H; Q\mathbf{d}_{H,k}; \mathbf{X}), \qquad \forall Q \in \mathrm{Orth}^+. \tag{2.10}$$

Hence, the stresses are not all independent, and we can imagine to express, for instance, $\boldsymbol{\tau}^H$ in terms of $\boldsymbol{\sigma}^i$ and $\boldsymbol{\eta}^{Hi}$, as done explicitly in the case of Cosserat media.

Moreover it is clear that $\boldsymbol{\sigma} \equiv \sigma^{ij}\mathbf{e}_i \otimes \mathbf{e}_j = \boldsymbol{\sigma}^i \otimes \mathbf{e}_j$ is a non symmetric tensor and it follows that also the corresponding Cauchy tensor is not symmetric.

3 Particular Models. I: Textile Fabrics

A particular case of the model presented in the previous section is the model of a textile fabric in which the vectorial microstructure is related to the microondulation of the welft and the warp.

We want to focus our attention to the theory of inextensible networks, in particular in the case in which a set of inextensible fibers forms a surface with bending stiffness and in which the effects due to the microundulations are taken into account, such that we can model the static behaviour of textile fabrics.

In 1986, Wang and Pipkin [16] formulated a theory of inextensible nets with bending stiffness. The resulting continuum theory is a special form of finite-deformation plate theory in which each fiber has a bending couple proportional to its curvature.

In 2001, a theory of bending and twisting effects in three-dimensional deformation of an inextensible network is presented by Luo and Steigmann [7]. They derive the Euler-Lagrange equations and boundary conditions by using the minimum-energy principle. (A simplified version of these equations represents the equilibrium equations obtained by Wang and Pipkin [16].)

In 2008, G. Indelicato [4, 5, 6] developed a more general model, which encompass the two previous models and introduced the microoundulation as well.

The aim of this work consists of finding the equations that express the effect of the shearing, the bending and also the microundulations of the fibers on the deformation of the sheet. Field equations are obtained via a variational principle. The strain energy density is written in an additive form, such that the contributions due to shearing, bending effects and microundulations are taken into account separately.

3.1 *Inextensible Fibers, Constitutive Hypotheses*

We consider two families of inextensible fibers forming a surface that initially lies in a region B of the (x,y)-plane. We assume that initially the first family of fibers, D_1, stays parallel to the x axis and that the second family of fibers, D_2, stays parallel to the y axis. We suppose that fibers are continuously distributed so that every line x=constant or y=constant in B is regarded as a fiber. Each fiber is permanently identified by its initial coordinate, x or y. We suppose that cross-sections of each fiber remain plane, suffer no strain, and are normal to the fiber in every configuration (Bernoulli-Euler hypotheses).

We denote the position in the current configuration with r(x,y), namely the point of the fibers that initially lies in the position (x,y) moves to the place r(x,y) in three-dimensional space.

Let

$$D_1 = \frac{\partial r}{\partial x} = r_{,x} \qquad D_2 = \frac{\partial r}{\partial y} = r_{,y} \tag{3.1}$$

be the tangential vectors to the curve occupied by a fiber y=constant and x=constant, respectively, when the sheet is deformed. We postulate that no part of any fiber can change its length in any admissible deformation so the vectors D_1 and D_2 are unit tangent vectors [14]. Since x and y are the arc length of the D_1 and D_2 lines, Frenet's formulas allow us to attach to each fiber the normal vector n and the binormal vector b, hence for the fiber D_1 the triad $\{D_1, n_1, b_1\}$ satisfies:

$$\begin{cases} \dfrac{\partial D_1}{\partial x} = K_1 n_1 \\[2mm] \dfrac{\partial n_1}{\partial x} = -K_1 D_1 + \tau_1 b_1 \\[2mm] \dfrac{\partial b_1}{\partial x} = -\tau_1 n_1 \end{cases} \tag{3.2}$$

with K_1 the principal curvature and τ_1 the torsion of the D_1 line. Similarly, for the fiber D_2 we introduce the Frenet triad $\{D_2, n_2, b_2\}$.

The sets of fibers D_1 and D_2 are related through the angle of shear γ, that is defined by the relation

$$\gamma = D_1 \cdot D_2$$

this angle describes the local distortion of the sheet.

Moreover, we introduce the normal vector:

$$N = \frac{D_1 \times D_2}{|D_1 \times D_2|}.$$

On the other side, if we consider the textile fabric at the microlevel, we can recognize the microundulations due to the weawing of the warp and the welft. We suppose the microcurvature be constant along each ondulation.

Setting

$$\psi_1 = k_1/2,$$

with k_1 microcurvature of the first set of fibers, the local stretch in the direction of D_1 becomes:

$$\varepsilon_1 = 1 - \frac{\sin \psi_1}{\psi_1}.$$

We introduce the director

$$d_1 \equiv \varepsilon_1 D_1 \tag{3.3}$$

associated to the first set of fibers, (similarly for the set of fibers D_2 we introduce the director $d_2 \equiv \varepsilon_2 D_2$).

3.2   Field Equations

In the previous model, the strain energy density is supposed to be a function of:

$$r_{,h}, d_H, d_{H,j}, X,$$

since we are interested in the bending of the surface, the strain energy function must some how depend on the macrocurvature of the fibers (i.e. the second derivatives of the position vector). Consequently, we suppose the strain energy density to be a function:

$$W = W(r_{,x}; r_{,y}; r_{,xx}; r_{,yy}; d_1; d_2; d_{1,x}; d_{2,y}; x; y) =$$

$$= W\left(D_1; D_2; \frac{\partial D_1}{\partial x}; \frac{\partial D_2}{\partial y}; d_1; d_2; \frac{\partial d_1}{\partial x}; \frac{\partial d_2}{\partial y}; x; y\right). \tag{3.4}$$

Let us remark that W also depends on the second derivatives of $\mathbf{r}$ because we want to take in account higher order effects. We assign the strain energy density in an additive form, such that there is no coupling between shearing, bending and microundulations.

$$W = W_0(\boldsymbol{r}_{,x} \cdot \boldsymbol{r}_{,y}) + \tfrac{1}{2}\Gamma(\mathbf{r}_{,xx} \cdot \mathbf{r}_{,xx} + \mathbf{r}_{,yy} \cdot \mathbf{r}_{,yy}) +$$

$$+\frac{1}{2}(A_1\boldsymbol{d}_1 \cdot \boldsymbol{d}_1 + A_2\boldsymbol{d}_2 \cdot \boldsymbol{d}_2 + B_1\boldsymbol{d}_{1,x} \cdot \boldsymbol{d}_{1,x} + B_2\boldsymbol{d}_{2,y} \cdot \boldsymbol{d}_{2,y})$$
(3.5)

where Γ is the same positive constant for all the fibers, namely the stiffness coefficient.

The energy component W_0 is due to the shearing stress, hence it can be assumed to depend on the angle between the fibers D_1 and D_2, namely W_0 is a function of $\boldsymbol{D}_1 \cdot \boldsymbol{D}_2$ only; the second component of the strain energy is associated to bending, it is a quadratic form in the fiber macrocurvatures; the third component of the strain energy is associated to microondulation and local macrostretch. Expression (3.5) can be written in the following explicit form:

$$W = W(\sin\gamma; K_1; K_2; \varepsilon_1; \varepsilon_2; \varepsilon_{1,x}; \varepsilon_{2,y}) =$$

$$= W_0(\sin\gamma) + \frac{1}{2}\Gamma[(k_1)^2 + (k_2)^2] + \frac{1}{2}(A_1(\varepsilon_1)^2 + A_2(\varepsilon_2)^2) +$$
(3.6)

$$+\frac{1}{2}B_1[(\varepsilon_{1,x})^2 + (\varepsilon_1 K_1)^2] + \frac{1}{2}B_2[(\varepsilon_{2,y})^2 + (\varepsilon_2 K_2)^2].$$

The kinetic energy reduced into diagonal form is:

$$T = \frac{1}{2}\rho\left(\dot{\mathbf{r}} \cdot \dot{\mathbf{r}} + I^{HK}\dot{\mathbf{d}}_H \cdot \dot{\mathbf{d}}_K\right)$$
(3.7)

with ρ the body mass density and I^{HK} is a kind of inertia tensor.

The total energy is given by:

$$E = \int_S (T - W)dA$$

with S the surface formed by the inextensible fibers.

Leaving apart the problem of the boundary conditions, the field equations can be derived via a variational principle as the Euler-Lagrange equations of the functional:

$$\mathcal{E} = \int_{t_0}^{t_1}\left[\int_S (T - W - (\lambda_1\boldsymbol{D}_1 \cdot \boldsymbol{D}_1 - 1) - (\lambda_2\boldsymbol{D}_2 \cdot \boldsymbol{D}_2 - 1))dA\right]dt \quad (3.8)$$

with λ_i the Lagrange multipliers associated to the constrain of inextensibility of the two sets of fibers:

$$\begin{cases} \boldsymbol{D}_1 \cdot \boldsymbol{D}_1 = 1 \\[2mm] \boldsymbol{D}_2 \cdot \boldsymbol{D}_2 = 1. \end{cases} \qquad (3.9)$$

since $\boldsymbol{D}_i \equiv \boldsymbol{r}_{,i}$, the Euler-Lagrange equations read:

$$\begin{cases} -\left(\dfrac{\partial W}{\partial \boldsymbol{r}_{,ih}}\right)_{,ih} + \left(\dfrac{\partial W}{\partial \boldsymbol{r}_{,i}}\right)_{,i} - (\lambda_i \boldsymbol{r}_{,i})_{,i} = \dfrac{d}{dt}\dfrac{\partial T}{\partial \dot{\boldsymbol{r}}} \\[4mm] \left(\dfrac{\partial W}{\partial \boldsymbol{d}_{H,i}}\right)_{,i} - \dfrac{\partial W}{\partial \boldsymbol{d}_H} = \dfrac{d}{dt}\dfrac{\partial T}{\partial \dot{\boldsymbol{d}}_H}. \end{cases} \qquad (3.10)$$

Using the expression (3.5) for the strain energy functions, recalling (3.1), (3.2), (3.3), the field equations become:

$$\begin{cases} -\Gamma\left[(K_1 \boldsymbol{n}_1)_{,xx} + (K_2 \boldsymbol{n}_2)_{,yy}\right] + \left(\dfrac{dW_0}{d\sin\gamma}\right)_{,x} \boldsymbol{D}_2 + \dfrac{dW_0}{d\sin\gamma}\dfrac{\partial \boldsymbol{D}_2}{\partial x} + \\[4mm] \quad + \left(\dfrac{dW_0}{d\sin\gamma}\right)_{,y} \boldsymbol{D}_1 + \dfrac{dW_0}{d\sin\gamma}\dfrac{\partial \boldsymbol{D}_1}{\partial y} + (\lambda_1 \boldsymbol{D}_1)_{,x} + (\lambda_2 \boldsymbol{D}_2)_{,y} = \rho\ddot{\boldsymbol{r}} \\[5mm] \left[B_1 \dfrac{\partial \varepsilon_1}{\partial x}\boldsymbol{D}_1 + \varepsilon_1 K_1 \boldsymbol{n}_1\right] - A_1 \boldsymbol{d}_1 = I^{11}\ddot{\boldsymbol{d}}_1 + I^{21}\ddot{\boldsymbol{d}}_2 \\[5mm] \left[B_2 \dfrac{\partial \varepsilon_2}{\partial y}\boldsymbol{D}_2 + \varepsilon_2 K_2 \boldsymbol{n}_2\right] - A_2 \boldsymbol{d}_2 = I^{12}\ddot{\boldsymbol{d}}_1 + I^{22}\ddot{\boldsymbol{d}}_2. \end{cases} \qquad (3.11)$$

Equations $(3.11)_2$,$(3.11)_3$ relate the microundulations with the local stretch of the fibers namely, the local stretch of the surface in the two directions $\boldsymbol{D}_1$ and $\boldsymbol{D}_2$.

The equations found by Wang and Pipkin in [16], and also by Luo and Steigmann in [7] for the case without twist, read:

$$\begin{cases} \boldsymbol{F}_1 = T_1 \boldsymbol{D}_1 + \dfrac{d\,W_0}{d\,\sin\gamma}\boldsymbol{D}_2 - \Gamma \boldsymbol{D}_{1,xx} \\[6mm] \boldsymbol{F}_2 = T_2 \boldsymbol{D}_2 + \dfrac{d\,W_0}{d\,\sin\gamma}\boldsymbol{D}_1 - \Gamma \boldsymbol{D}_{2,yy} \\[6mm] \dfrac{\partial \boldsymbol{F}_1}{\partial x} + \dfrac{\partial \boldsymbol{F}_2}{\partial y} + \boldsymbol{f} = \boldsymbol{0}. \end{cases} \qquad (3.12)$$

Substituting $(3.12)_1$, $(3.12)_2$ in $(3.12)_3$ we obtain:

$$\frac{\partial(T_1 \boldsymbol{D}_1)}{\partial x} + \left(\frac{\mathrm{d}W_0}{\mathrm{d}\sin\gamma}\right)_{,x} \boldsymbol{D}_2 + \frac{\mathrm{d}W_0}{\mathrm{d}\sin\gamma}\frac{\partial \boldsymbol{D}_2}{\partial x} - \Gamma\left(\frac{\partial \boldsymbol{D}_{1,xx}}{\partial x}\right) +$$

$$\frac{\partial(T_2 \boldsymbol{D}_2)}{\partial y} + \left(\frac{\mathrm{d}W_0}{\mathrm{d}\sin\gamma}\right)_{,y} \boldsymbol{D}_1 + \frac{\mathrm{d}W_0}{\mathrm{d}\sin\gamma}\frac{\partial \boldsymbol{D}_1}{\partial y} - \Gamma\left(\frac{\partial \boldsymbol{D}_{2,yy}}{\partial y}\right) + \boldsymbol{f} = \boldsymbol{0}$$

$$(3.13)$$

with: $T_1 = \lambda_1$ and $T_2 = \lambda_2$ reactions to the constraints of fiber inextensibility. Equation (3.13) is exactly equation $(3.11)_1$ in the static case, with $\boldsymbol{f} = \boldsymbol{0}$.

4 Particular Models. II: Plane Granular Media

In this Section, we study a simpler dissipative model of plane granular media.

If the body is a 2-D solid and its configuration at any time is a domain contained in $\mathbb{R}^2$, we can choose an orthonormal spatial basis $\{\mathbf{e}_h\}$, $h = 1, 2$ and a material basis $\{\mathbf{g}_h\}$, where $\mathbf{g}_h = \mathbf{r}_{,h}$, hence write $\mathbf{r} = x^h(X^k, t)\mathbf{e}_h$; the functions $x^h(X^k, t)$ have the meaning of deformation function components. Latin indices take the values 1,2.

If the directors $\mathbf{d}_\alpha$ are written in the spatial basis $\{\mathbf{e}_h\}$ by means of the angle of rotation θ, such that

$$\mathbf{d} = \mathbf{d}_1 = \cos\theta\mathbf{e}_1 + \sin\theta\mathbf{e}_2$$
$$\boldsymbol{\nu} = \mathbf{d}_2 = -\sin\theta\mathbf{e}_1 + \cos\theta\mathbf{e}_2$$

the conditions $\mathbf{d}_\alpha \cdot \mathbf{d}_\beta = \delta_{\alpha\beta}$ are automatically satisfied. Physically, we interpret $\mathbf{d}$ as the kinematical characterization of the grain and it is fully determined by the scalar function $\theta(X^k, t)$.

4.1 The Field Equation

The kinetic energy density reads:

$$T = \frac{1}{2}\left[\rho\left(u_t^2 + v_t^2\right) + I\theta_t^2\right]$$

The strain energy density is choosen in the form:

$$\mathcal{W} = \frac{1}{2}\alpha\left(u_x^2 + v_x^2\right) + \frac{1}{2}\beta\left(u_y^2 + v_y^2\right) + \frac{1}{6}\gamma(u_x^3 + u_y^3 + v_x^3 + v_y^3)$$

$$+ \frac{1}{2}\gamma(u_x^2 v_x + u_y^2 v_y + v_x^2 u_x + v_y^2 u_y) - A\theta\left(u_x + u_y + v_x + v_y\right)$$

$$+ \frac{1}{2}B\theta^2 + \frac{1}{2}C\left(\theta_x^2 + \theta_y^2\right) + \frac{1}{3}D\left(\theta_x^3 + \theta_y^3\right)$$

The coefficients α, β, γ, A, B, C, D are constants depending on the material properties of the granular solid.

We introduce a new variable $w = u + v$ such that the field equation read:

$$\begin{cases} \rho w_{tt} = \alpha w_{xx} + \beta w_{yy} - 2A(\theta_x + \theta_y) + \gamma\left[(w_x^2)_x + (w_y^2)_y\right] \\[2ex] I\theta_{tt} = C(\theta_{xx} + \theta_{yy}) + D\left[(\theta_x^2)_x + (\theta_y^2)_y\right] + A(w_x + w_y) - B\theta - F(\theta_x + \theta_y)_t \end{cases}$$

Let remark that the last term $-F(\theta_x + \theta_y)_t$ represent the dissipation due to the friction among particles.

From the second equation of the previous system, we obtain the following expression for θ:

$$\theta = \frac{1}{B}\left\{- I\theta_{tt} + C(\theta_{xx} + \theta_{yy}) + D\left[(\theta_x^2)_x + (\theta_y^2)_y\right] + A(w_x + w_y) - F(\theta_x + \theta_y)_t\right\}$$

We consider the dimensionless form of the previous system, and we apply the slaving principle (for all passages see [1]). For further analysis the dimensionless variables are introduced (note that θ is already dimensionless)

$$W = \frac{w}{W_0}, \quad X = \frac{x}{L}, \quad Y = \frac{y}{L}, \quad T = \frac{c_0^2}{L}t$$

where c_0^2, W_0, L are physically meaningful certain constants (velocity, intensity and wavelenght of the initial excitation). We also need a scale for the microstructure l; accordingly, two dimensionless parameters can be introduced

$$\begin{cases} \delta \sim \left(\dfrac{l}{L}\right)^2 \quad \text{characterizing the ratio between the} \\[2ex] \qquad\qquad\qquad \text{microstructure and the wave lenght;} \\[2ex] \epsilon \sim \left(\dfrac{W_0}{L}\right) \quad \text{accounting for elastic strain} \end{cases}$$

where δ has the relevant meaning of a characteristic length.

We suppose $I = \rho l^2 I^*$, $C = l^2 C^*$, $D = l^2 D^*$, $F = l^2 F^*$ where I^* is dimensionless and C^* D^* and F^* have the dimension of a stress.

We then obtain:

$$\begin{cases} W_{TT} = \dfrac{1}{\rho c_0^4}(\alpha W_{XX} + \beta W_{YY}) - \dfrac{2A}{\rho c_0^4 \epsilon}(\theta_X + \theta_Y) + \epsilon\gamma\left[(W_X^2)_X + (W_Y^2)_Y\right] \\[3ex] \theta = \dfrac{\epsilon A}{B}(W_X + W_Y) + \dfrac{\delta}{B}\left[C^*(\theta_{XX} + \theta_{YY}) + \dfrac{D^*}{L}\left[(\theta_X^2)_X + (\theta_Y^2)_Y\right]\right] + \\[3ex] \qquad\qquad - \dfrac{\delta}{B}\left[\rho c_0^4 I^*\theta_{TT} + c_0 F^*(\theta_{XT} + \theta_{YT})\right] \end{cases}$$

If we consider the expansion in terms of the caracteristic lenght δ:

$$\theta = \theta_0 + \delta\theta_1 + \ldots \simeq \frac{\epsilon A}{B}(W_X + W_Y) + \frac{\delta\epsilon A}{B^2}\left\{ C^* \left[(W_X + W_Y)_{XX} + (W_X + W_Y)_{YY}\right]\right.$$

$$\left. - \rho c_0^4 I^*(W_X + W_Y)_{TT} - c_0 F^* \left[(W_X + W_Y)_{XT} + (W_X + W_Y)_{YT}\right] \right\}$$

we obtain the governing equation:

$$W_{TT} = \frac{1}{\rho c_0^4}(\alpha W_{XX} + \beta W_{YY}) - \frac{2A^2}{\rho c_0^4 B}(W_{XX} + 2W_{XY} + W_{YY}) +$$

$$- \frac{2\delta A^2 C^*}{\rho c_0^4 B^2}(W_{XXXX} + 2W_{XXXY} + 2W_{XXYY} + 2W_{XYYY} + W_{YYYY}) +$$

$$+ \frac{2\delta A^2}{B^2}\left[I^*(W_{XX} + 2W_{XY} + W_{YY})_{TT} + c_0 G^*(W_{XXX} + 3W_{XXY} + 3W_{XYY} + W_{YYY})_T\right] +$$

$$+ \epsilon\gamma\left[(W_X^2)_X + (W_Y^2)_Y\right]$$

These equations are suitable to study non linear wave propagations in such solids, (see for instance [1], [3], [12]), but we choose to stop here our exposition. Following the same path, changing the strain energy fuction, we can derive other PDE's, i.e. sixth order PDE's which can be reduced to a fourth order principal ODE, following the slaving principle (see for instance [2]) and a method introduced by A. Samsonov [15]. Hence it seems feasible to study non linear waves and solitons propagation and, from another point of you, to establish a hierarchy of waves (see [3]), which can be useful to detect the influence of the different parameters on the wave propagation. In general, the possibility to obtain Euler - Lagrange equations as field equations for solids with microstrucutres as proved in this paper gives rise to a theory that can encompass many different particular models of solids, and fuids too, where micro structures and complex inner structures have a relevant role.

References

1. Casasso, A., Pastrone, F.: Nonlinear Waves Motion in Complex Elastic Structures. Rend. Circ. Palermo. Serie II 78(suppl.), 45–58 (2006)
2. Engelbrecht, J.: Nonlinear wave dynamics. Complexity and simplicity. Kluwer, Dordrecht (1997)
3. Engelbrecht, J., Pastrone, F., Braun, M., Berezovski, A.: Hierarchy of waves in non-classical materials. In: Delsanto, P.P. (ed.) The universality of nonclassical nonlinearity, with applications to NDE and Ultrasonics. Springer, New York (2006)

4. Indelicato, G.: Mechanical models for 2D fiber networks and textiles, PhD Thesis, Torino (2008)

5. Indelicato, G.: Inextensible networks with bending and twisting effects. Rend. Sem. Mat. Un. Pol. Torino 65(2), 261–268 (2007)

6. Indelicato, G.: The influence of the twist of individual fibers in 2D fibered networks. Intl. J. Solids and Structures 46(3-4), 912–922 (2009)

7. Luo, C., Steigmann, D.J.: Bending and Twisting Effects in the Three-Dimentional Finite Deformations of an Inextensible Network. In: Advances in the Mechanics of Plates and Shells, pp. 213–228. Kluwer Academic Publishers, Dordrecht (2001)

8. Maugin, G.A.: Nonlinear Waves in Elastic Crystal. Oxford University Press, UK (1999)

9. Mindlin, R.D.: Microstructure in linear elasticity. Arch. Ratl. Mech. and Anal. 1, 51–78 (1964)

10. Pastrone, F.: Mathematical Models of Microstructured Solids. Lect. Notes Mech. 4/04 Tallinn Tech. Univ. Tallinn (2004)

11. Pastrone, F.: Waves Propagation in Microstructured Solids. Mathematics and Mechanics of Solids 10, 349–357 (2005)

12. Pastrone, F.: Nonlinearity and complexity in Elastic Wave Motion. In: Delsanto, P.P. (ed.) The universality of nonclassical nonlinearity, with applications to NDE and Ultrasonics. Springer, New York (2006)

13. Pastrone, F.: Microstructures and granular media. Rend. Sem. Univ. Pol. Torino 65(2) (2007)

14. Pipkin, A.C.: Some Developments in the Theory of Inextensible Networks. Quart. Appl. Math. 38, 343–355 (1980)

15. Samsonov, A.M.: Strain soliton in solids and how to construct them. Chapman & Hall/CRC, New York (2000)

16. Wang, W.B., Pipkin, A.C.: Inextensible Networks with Bending Stiffness. Q. Jl. Mech. Appl. Math. 39, 359–434 (1986)

Numerical Simulation of Interaction of Solitons and Solitary Waves in Granular Materials

Andrus Salupere and Lauri Ilison

Abstract. A hierarchical Korteweg–de Vries type evolution equation is used for modelling of wave propagation in dilatant granular materials. The model equation is integrated numerically under sech^2-type initial conditions using the discrete Fourier transform based pseudospectral method. In our previous papers we have shown that depending on values of material parameters five different solution types can be detected. In all cases one component of the solution is a solitary wave or an ensemble of solitary waves (solitons) that can propagate at constatnt speed and amplitude and in cases of ensembles interact (almost) elastically. In the present paper additional numerical experiments for simulation of interactions between different soliton ensembles, single solitons and solitary waves are carried out and analysed.

1 Introduction

Many physical and technological applications deal with nonlinear wave propagation in continuous media with microstructure. For that reason attention to corresponding studies in recent years has been increased (see e.g. [3, 4, 15] and references therein). Granular materials are an example of such microstructured materials [1, 7, 16, 17]. In the present paper wave propagation in granular materials is modelled by the hierarchical Korteweg–de Vries (HKdV) equation [1]

Andrus Salupere
Centre for Nonlinear Studies, Institute of Cybernetics at Tallinn University of Technology, and Department of Mechanics, Tallinn University of Technology, Akadeemia tee 21, 12618 Tallinn, Estonia
e-mail: salupere@ioc.ee

Lauri Ilison
Centre for Nonlinear Studies, Institute of Cybernetics at Tallinn University of Technology, and Department of Mechanics, Tallinn University of Technology, Akadeemia tee 21, 12618 Tallinn, Estonia
e-mail: lauri@cens.ioc.ee

J.-F. Ganghoffer, F. Pastrone (Eds.): Mech. of Microstru. Solids 2, LNACM 50, pp. 21–28.
springerlink.com
© Springer-Verlag Berlin Heidelberg 2010

$$\frac{\partial u}{\partial t} + u\frac{\partial u}{\partial x} + \alpha_1\frac{\partial^3 u}{\partial x^3} + b\frac{\partial^2}{\partial x^2}\left(\frac{\partial u}{\partial t} + u\frac{\partial u}{\partial x} + \alpha_2\frac{\partial^3 u}{\partial x^3}\right) = 0. \tag{1}$$

Here α_1 and α_2 are macro- and microlevel dispersion parameters, respectively and b is the microstructure parameter that involves the ratio of the grain size and the wavelength. Equation (1) consists of two KdV operators: the first describes the motion in the macrostructure and the second (in the brackets) — the motion in the microstructure. Equation (1) is clearly hierarchical in the Whitham's sense [2] — if parameter b is small then the influence of the microstructure can be neglected and the wave "feels" only macrostructure. If, however, parameter b is large, then only the influence of the microstructure "is felt" by the wave. The limiting case ($\beta = 0$) results in the standard KdV equation with standard soliton solutions.

In our previous studies the main goal was to simulate emergence of solitons and soliton ensembles from a single sech^2-type initial pulse [9, 10, 12]. Based on the analysis of numerical results we have demonstrated the existence of five different solution types: (i) single KdV soliton, (ii) KdV soliton ensemble, (iii) KdV soliton ensemble a with weak tail, (iv) soliton with a strong tail, and (v) soliton with a tail and wave packet. The main aim of this paper is to simulate and analyse interactions between the solitary waves that emerge from different sech^2-type initial pulses.

2   Statement of the Problem and Numerical Technique

We have demonstrated in [9, 10], that (i) the solution type is determined by the values of material parameters α_1, α_2 and β and it does not depend on the value of the amplitude (height) of the initial pulse; (ii) two initial pulses having different amplitudes generate solitons or solitary waves that correspond to the same solution type but propagate at different speeds and therefore can interact during the propagation.

In order to simulate interactions we shall use here an initial condition that consists of two different amplitude sech^2-type localised solitary waves which are shifted with the respect to $x = 0$ by 16π and 48π respectively:

$$u(x,0) = A_1\mathrm{sech}^2\frac{x - 16\pi}{\delta_1} + A_2\mathrm{sech}^2\frac{x - 48\pi}{\delta_2},$$
$$\delta_1 = \sqrt{\frac{12\alpha_1}{A_1}}, \qquad \delta_2 = \sqrt{\frac{12\alpha_1}{A_2}}. \tag{2}$$

Here A_1 is the amplitude of the left hand side initial pulse and A_2 is the amplitude of the right hand side one, $0 \leq x < 64\pi$, and δ_1 and δ_2 are the widths of the initial pulses.

The goals of the present paper are:

(i) to simulate numerically interactions between

   a. two single KdV solitons (the first solution type in [9, 10, 12]);

   b. solitons from different KdV soliton ensembles (the second and the third solution types in [9, 10, 12]);

   c. two solitons (solitary waves) with strong tails (the fourth solution type in [9, 10, 12]);

(ii) to analyse the character of interactions in terms of solitons, i.e. to understand whether solitary waves that emerge from different initial pulses interact elastically or not.

For numerical integration of the HKdV equation the pseudospectral method (PsM) [5, 14, 20] is applied. In a nutshell, the idea of the PsM is to approximate space derivatives by a certain global method — reducing thereby partial differential equation to ordinary differential equation (ODE) — and to apply a certain ODE solver for integration with respect to the time variable. In the present paper space derivatives are found making use of the discrete Fourier transform (DFT). Calculations are carried out using SciPy package [13]: for DFT the FFTW [6] library and for ODE solver the F2PY [18] generated Python interface to ODEPACK Fortran code [8] is used.

3   Results and Discussion

The HKdV equation (1) was integrated numerically under initial conditions (2) and periodic boundary conditions

$$u(x+64k\pi,t) = u(x,t), \qquad k = \pm1,\pm2,\pm3,\ldots. \tag{3}$$

The values of dispersion parameters α_1, α_2 and microstructure parameter β have been selected according to the solution types defined in [9, 10, 12]. The number of space grid points $n = 4096$ and the length of the time interval $t_f = 100$.

3.1   *Interactions of Single KdV Solitons*

The first solution type is called the single KdV soliton and it appears if dispersion parameters $\alpha_1 = \alpha_2$. In this case the initial sech^2-pulse propagates at a constant speed and a constant amplitude [9, 10, 12]. Here we simulate interactions between two initial pulses that have different amplitudes ($A_1 = 15$ and $A_2 = 5$) and therefore they propagate at different speeds. Analysis of numerical results demonstrate clearly that interactions between solitons are elastic as the solitons restore their amplitude (see Fig. 1) and speed after interactions. During the interaction solitons are phase shifted — higher amplitude soliton is shifted to the right and lower amplitude soliton to the left.

3.2   *Interactions of Solitons from KdV Soliton Ensembles*

In our previous paper [9] we found that it is quite conditional to distinguish between KdV soliton ensemble and KdV soliton ensemble with a weak tail, i.e., between the second and the third solution types. The tail is sometimes so weak, that it is

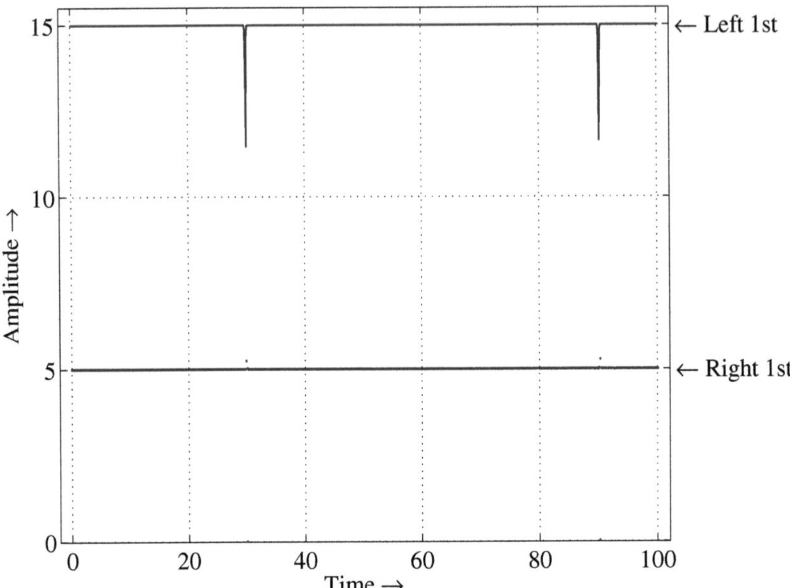

Fig. 1 Interactions of KdV solitons. Wave profile maxima against time for $\alpha_1 = \alpha_2 = 0.03$, $\beta = 0.0111, A_1 = 15, A_2 = 5$

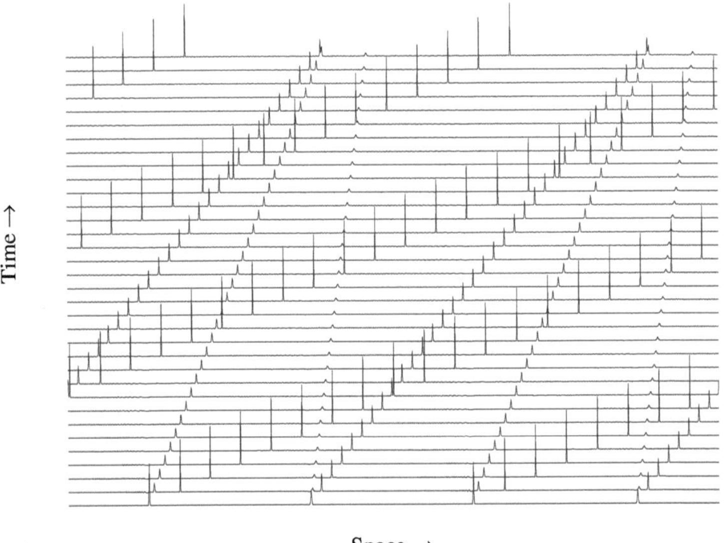

Fig. 2 Interactions of solitons from KdV soliton ensembles. Timeslice plot over two space periods for $\alpha_1 = 0.07, \alpha_2 = 0.03, \beta = 111.11, A_1 = 15, A_2 = 5$

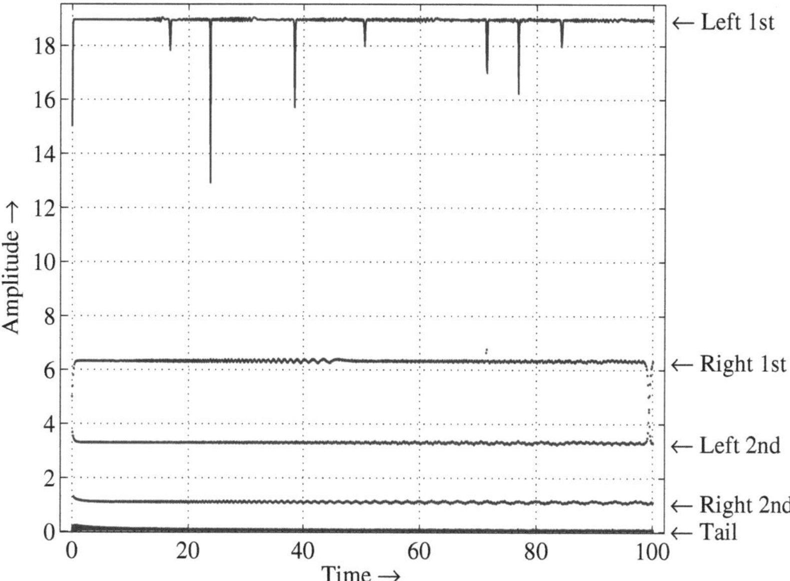

Fig. 3 Interactions of solitons from KdV soliton ensembles. Wave profile maxima against time for $\alpha_1 = 0.07$, $\alpha_2 = 0.03$, $\beta = 111.11$, $A_1 = 15$, $A_2 = 5$

practically indistinguishable by means of wave profile extrema as well as spectral quantities. For this reason we consider here these two types together.

As an example, we present interactions between solitons from two different ensembles that emerged from two initial pulses having amplitudes $A_1 = 15$ and $A_2 = 5$. Different amplitude solitons propagate at different speed and therefore interactions between them take place. One can trace here two type of interactions: (i) between solitons from different ensembles, and (ii) between solitons from the same ensemble. Time-slice plot (Fig. 2) and amplitude curves (Fig. 3) demonstrate that interactions of both type are practically elastic, i.e., solitons restore their speed and amplitude after interactions. Besides the soliton-soliton interactions all solitons interact with tails. However, as the tails are weak, they do not influence the behaviour of solitons essentially and their influence can be traced only in curves of wave profile maxima, where tails can cause small oscillations.

3.3 *Interactions of Solitons with Strong Tails*

In the present case two solitons and strong tails emerge from the initial excitation (2) (see Figs. 4 and 5). For this solution type the tail is considered to be strong, because it influences the behavior of emerged solitary waves essentially: (i) amplitudes of propagating solitons are not constant, but due to the influence of tails they oscillate with respect to a constant level remarkably (cf. Figs. 5 and 3); (ii) amplitudes of

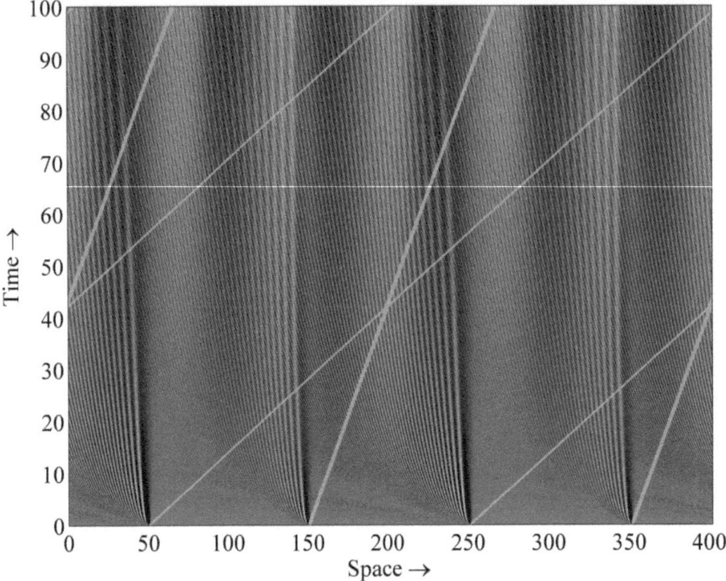

Fig. 4 Interactions of solitons with strong tails. Pseudocolor plot over two space periods for $\alpha_1 = 0.03$, $\alpha_2 = 0.07$, $\beta = 111.11$, $A_1 = 15$, $A_2 = 5$

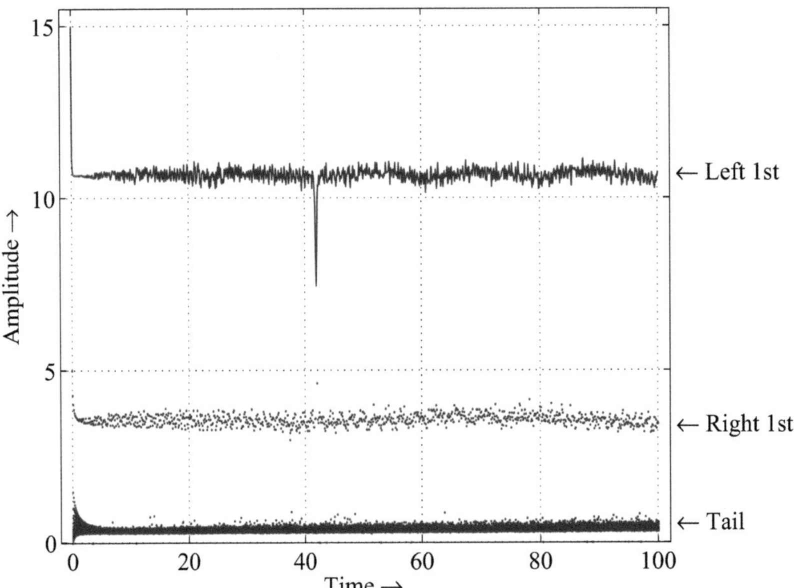

Fig. 5 Interactions of solitons with strong tails. Wave profiles amplitude plot over two space periods for $\alpha_1 = 0.03$, $\alpha_2 = 0.07$, $\beta = 111.11$, $A_1 = 15$, $A_2 = 5$

the propagating solitary waves are lower than the amplitudes of the initial ones (see Fig. 5). Such a phenomenon — the shape of the initial solitary wave is altered during propagation — is called selection (see [2, 19] for details). It modifies the shape of the initial condition in a way to be more appropriate to the real solution of the equation.

Analysis of the behaviour of trajectories and amplitudes of selected solitary waves demonstrate that the interaction is nearly elastic, i.e., the trajectories are phase-shifted during interaction and after the interaction both solitons almost restore their speeds and amplitudes (Figs. 4 and 5). Therefore these solitary waves can be called solitons.

4 Conclusions

In the present paper interactions of solitary waves in media governed by HKdV equation (1) are examined. The model equation is integrated numerically under initial condition (2) and periodic boundary condition (3). The analysis of numerical results demonstrate that emerged solitary waves interact elastically or nearly elastically in all considered cases. More numerical examples can be found in unpublished Research Report [11], where interactions between two solitary waves with tails and wave packets (the fifth solution type in [9, 10, 12]) are also considered. The latter case is more complicated than these considered here. We have found that for this solution type (i) interacting solitary waves emerge only in few cases, and (ii) if interactions take place then they are not elastic by means of solitons.

Acknowledgements. Authors of this paper thank Professor Jüri Engelbrecht for helpful discussions and senior researcher Pearu Peterson for Python related scientific software developments and assistance. The research is supported by Estonian Science Foundation Grant No 7035.

References

1. Cataldo, G., Oliveri, F.: Nonlinear seismic waves: A model for site effects. Int. J. Nonlinear Mech. 34, 457–468 (1999)
2. Christov, C., Velarde, M.: Dissipative solitons. Physica D 86, 323–347 (1995)
3. Engelbrecht, J., Berezovski, A., Pastrone, F., Braun, M.: Waves in microstructured materials and dispersion. Phil. Mag. 85, 4127–4141 (2005)
4. Erofeev, V.I.: Wave Processes in Solids with Microstructure. World Scientific, Singapore (2003)
5. Fornberg, B.: Practical Guide to Pseudospectral Methods (1998)
6. Frigo, M., Johnson, S.G.: The design and implementation of FFTW3. Proceedings of the IEEE 93 (2), 216–231 (2005)
7. Giovine, P., Oliveri, F.: Dynamics and wave propagation in dilatant granular materials. Meccanica 30, 341–357 (1995)
8. Hindmarsh, A.C.: ODEPACK, a systematized collection of ODE solvers. In: Stepleman, R.S., et al. (eds.) Scientific Computing, pp. 55–64. North-Holland, Amsterdam (1983)

9. Ilison, L., Salupere, A.: Propagation of $sech^2$-type solitary waves in hierarchical KdV-type systems. In: Mathematics and Computers in Simulation (submitted)
10. Ilison, L., Salupere, A.: Propagation of localised perturbations in granular materials. Research Report Mech 287/07, Institute of Cybernetics at Tallinn University of Technology (2007)
11. Ilison, L., Salupere, A.: Interactions of solitary waves in hierarchical KdV-type system. Research Report Mech 291/08, Institute of Cybernetics at Tallinn University of Technology (2008)
12. Ilison, L., Salupere, A., Peterson, P.: On the propagation of localised perturbations in media with microstructure. Proc. Estonian Acad. Sci. Phys. Math. 56(2), 84–92 (2007)
13. Jones, E., Oliphant, T., Peterson, P., et al.: SciPy: Open source scientific tools for Python (2007), http://www.scipy.org
14. Kreiss, H.O., Oliger, J.: Comparison of accurate methods for the integration of hyperbolic equations. Tellus 30, 341–357 (1972)
15. Maugin, G.: Nonlinear Waves in Elastic Crystals. Oxford Univ. Press, Oxford (1999)
16. Oliveri, F.: Wave propagation in granular materials as continua with microstructure: Application to seismic waves in a sediment filled site. Rendiconti Circolo Matematico di Palermo 45, 487–499 (1996)
17. Oliveri, F., Speciale, M.P.: Wave hierarchies in continua with scalar microstructure in the plane and spherical symmetry. Computers and Mathematics with Applications 55, 285–298 (2008)
18. Peterson, P.: F2PY: Fortran to Python interface generator (2005), http://cens.ioc.ee/projects/f2py2e/
19. Porubov, A.V.: Amplification of Nonlinear Strain Waves in Solids. World Scientific, Singapore (2003)
20. Salupere, A.: On the application of the pseudospectral method for solving the Korteweg–de Vries equation. Proc. Estonian Acad. Sci. Phys. Math. 44(1), 73–87 (1995)

Multi-scale Modelling of Fracture in Open-Cell Metal Foams

K.R. Mangipudi and P.R. Onck

1 Introduction

Metal foams possess attractive mechanical properties like high stiffness to weight ratio. When used to build light-weight structures they require a good combination of strength and ductility. They are ductile under compression but rather brittle in tension with a few percent of overall strain to fracture. Second-phase particles, grain boundary precipitates and inclusions are often associated with the knock-down of the ductility [?]. Through a heat treatment the microstructure can be changed, however, this also changes the associated yield stress and hardening behaviour of the strut material. How this will affect the overall behaviour depends sensitively on the foam's cellular architecture, e.g. the cell size and shape distribution, the cross-sectional geometry of the strut, and its relative density. The goal of this work is to study these dependencies using a multi-scale modelling framework that takes all these ingredients into account. In this paper, we present the combined effect of the solid material strain hardening and the relative density on the initiation and accumulation of damage and overall strength of the structure.

2 Multi-scale Model

Random Voronoi structures in two dimensions are used to describe the structure of open-cell metal foams (Fig. 1). The mechanical behaviour of each material point in the strut is characterized by a uniaxial tensile curve featuring linear elasticity, power law hardening and softening due to damage. Individual struts are discretized with Euler-Bernoulli beam elements. In the following section, the

K.R. Mangipudi and P.R. Onck
Zernike Institute for Advanced Materials, Department of Applied Physics,
University of Groningen, The Netherlands
e-mail: `k.r.mangipudi@rug.nl`

J.-F. Ganghoffer, F. Pastrone (Eds.): Mech. of Microstru. Solids 2, LNACM 50, pp. 29–36.
springerlink.com © Springer-Verlag Berlin Heidelberg 2010

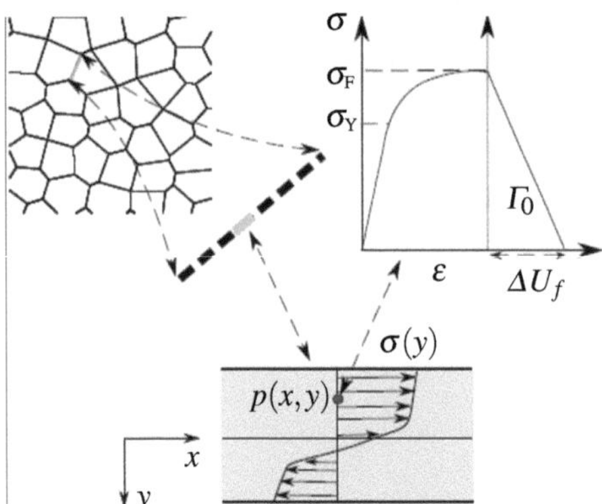

Fig. 1 Information flow between different length scales

governing equations and the framework for the plasticity and fracture is presented within a linear Finite Element (FE) setup. An updated Lagrange formulation is adopted. It should be emphasized here that the viscoplastic framework, which will be described later, is used only for numerical reasons; we are not dealing with physical time-dependent processes.

2.1 Finite Element Equations

We consider the strut to be modelled as an initially straight Euler-Bernoulli (EB) beam. In a structure consisting of many curved struts at different orientations, the beam axis is discretized into linear segments with appropriately defined local coordinate systems. We define the local coordinate system of the beam with its x-axis aligned with the bam axis represented by $x \in [0,L]$. Its deformations are described by the axial and transverse displacements of its axis: $\mathbf{u} := \{u,v\}$. For small strains (though at arbitrary displacements), the axial strain $\bar{\varepsilon} = u'$ and the curvature $\kappa = v''$ are taken as the strain measures. The corresponding work conjugates are the axial force P and the bending moment M. Using a shorthand notation the strain and the stress measures are

$$\mathbf{e} = \{\bar{\varepsilon}, \kappa\} \quad \text{and} \quad \mathbf{s} = \{P, M\}. \tag{1}$$

We decompose the total strain rate into an elastic and viscoplastic part,

$$\dot{\mathbf{e}} = \dot{\mathbf{e}}^e + \dot{\mathbf{e}}^{vp}, \tag{2}$$

and the linear elastic constitutive relation written as $\mathbf{s} = \mathsf{C}\mathbf{e}^e$, where C is the effective stiffness matrix. The rate form of principle of virtual work describing the equilibrium of the beam is given by

$$\int_x \left[\delta\mathbf{e}^T \dot{\mathbf{s}} + \dot{\delta\mathbf{e}}^T \mathbf{s} \right] dx = \int_x \delta\mathbf{u}^T \mathbf{f}_{ext}\, dx, \tag{3}$$

for all admissible variations $\delta \mathbf{u}$ and for the given external loading $\dot{\mathbf{f}}$. Making use of Eq. 2 along with the constitutive relation and by setting $\dot{\mathbf{s}}^* = \mathbf{C}\dot{\mathbf{e}}^{vp}$, we can rewrite Eq. 3 as

$$\int_x \left[\delta \mathbf{e}^T \mathbf{C}\dot{\mathbf{e}} + \dot{\delta \mathbf{e}}^T \mathbf{s} \right] dx = \int_x \delta \mathbf{u}^T \dot{\mathbf{f}}_{ext}\, dx + \int_x \delta \mathbf{e}^T \dot{\mathbf{s}}^*. \tag{4}$$

Using the standard beam interpolation functions for the displacements as a function of the nodal displacements and rotations, the following updated Lagrangian finite element equations are obtained:

$$(\mathbf{K}_M + \mathbf{K}_G)\dot{\mathbf{U}} = \dot{\mathbf{F}}_{ext} + \dot{\mathbf{F}}^*, \tag{5}$$

where $\mathbf{K}_M$ and $\mathbf{K}_G$ are the material and geometric stiffness matrices respectively, $\dot{\mathbf{U}}$ is the rate of unknown nodal displacements and rotations and $\dot{\mathbf{F}}^*$ is the rate of viscoplastic forces.

Plasticity and Fracture

Plasticity is modelled in a similar way as nonlinear elasticity for beams by discretizing into fibers. Through the beam's depth m number of integration points are defined which can be interpreted as the fiber centers. For any fiber at $p(x,y)$ (see Fig. 1), the stresses and strains are given by

$$\dot{\varepsilon}_{xx}(y) = \dot{\bar{\varepsilon}} - y\dot{\kappa} \quad \text{and} \quad \dot{\sigma}_{xx}(y) = E^t(y)\dot{\varepsilon}_{xx}, \tag{6}$$

where $\dot{\bar{\varepsilon}}$ is the axial strain rate, $\dot{\kappa}$ is the curvature rate and E^t is the tangent modulus. These rates are integrated in time to store the fiber stress and strain. Yielding of any fiber occurs when the fiber stress exceeds the yield stress of the material. The elastic strain in a fiber after the yielding is neglected in comparison with the plastic strain increments. A power-law strain hardening of the form

$$\sigma = \sigma_Y \left(1 + \frac{E}{\sigma_Y}\varepsilon^p \right)^{N_s} \tag{7}$$

is used for the fibers. Here E is the Young's Modulus, N_s is the strain hardening exponent and σ_Y and ε_Y are the yield stress and yield strain respectively, ε^p is the accumulated plastic strain and ε_F is the fracture strain of the material. The tangent modulus $E^t(y)$ of the fiber is obtained from the uniaxial tensile curve of the solid material (Eq. 7). By integrating through the thickness of the beam, the effective stretching and bending stiffnesses of the beam are obtained.

The following viscoplastic relations for the axial strain rate and curvature rate can be derived for an EB beam [2]:

$$\dot{\bar{\varepsilon}}^{vp} = \dot{\varepsilon}_0 \left(\frac{P}{P_0} \right)^n \quad \text{and} \quad \dot{\kappa}^{vp} = \dot{\kappa}_0 \left(\frac{M}{M_0} \right)^n, \tag{8}$$

where $\dot{\varepsilon}_0$ and n are parameters, $P_0 = \sigma_0 t$, σ_0 is the reference stress, t is the beam thickness, M is the bending moment and M_0 is the reference moment. Taking Eq. 8 with $n \to \infty$ the viscoplastic formulation becomes rate-independent and can be used as a numeric algorithm so that the normal force P and moment M in the beam follow the reference force (P_0) and reference moment (M_0) given by any other constitutive relation.

When the strain at either of the extreme fibers reaches the critical fracture strain (ε_F) of the material (i.e. the strain at σ_F, see Fig. 1), damage is initiated which reduces the forces and moment to zero with the help of a damage parameter $D \in [0, 1]$. The damage parameter is formulated in terms of the fracture displacement jump ($\Delta u_f(y)$) in the fibers to give mesh-independency. The damage parameter D is defined in terms of a specific fracture energy (energy per unit area of the cross section), Γ_0,

$$D = \frac{1}{t\Delta U_f} \int_y \Delta u_f(y) dy, \tag{9}$$

where $\Delta U_f = 2\Gamma_0/\sigma_F$ (see Fig. 1). At the initiation of fracture, the normal force (P_{init}) and the moment (M_{init}) in the beam element are recorded. We assume linear softening for the normal force and moment according to

$$\begin{pmatrix} P_0 \\ M_0 \end{pmatrix} = (1 - D) \begin{pmatrix} P_{init} \\ M_{init} \end{pmatrix}. \tag{10}$$

During the fracture of an element, Eq. 10 provides the reference quantities for the viscoplastic framework to be used in Eq. 8. At the beginning of each time step, the vector of the rate of the nodal viscoplastic forces $\dot{\mathbf{F}}^*$ is computed based on the generalized stress state in the previous increment (e.g. P and M) and will be used in Eq. 5.

3 Results

Simulations were performed on Voronoi structures of 12×16 ($W \times H$) cells with a narrow cell size distribution and uniform strut thickness. The size of the structure is selected to converge for the plastic collapse stress. Every strut is discretized into 10 beam elements initially and each element is subsequently remeshed once into 10 elements when the critical fiber stress exceeds $0.85\sigma_Y$. Seven relative densities are considered between 0.04 and 0.24 and the for the N_s values of 0.05, 0.1, 0.17, 0.2 and 0.3 are used in the parametric study. Other inputs of the model are the Young's Modulus $E = 70$ GPa, the yield stress $\sigma_Y = 41$ MPa, the fracture strain $\varepsilon_F = 0.147$ and the specific fracture energy $\Gamma_0 = 20 \times 10^3$ N/m.

Fig. 2 shows a typical stress-strain curve. Four distinct regions can be observed: (i) a linear-elastic regime, (ii) a nonlinear transition regime (from point a to roughly point b) (iii) a nonlinear hardening regime (from point b to d) and (iv) an unloading regime (after point d). The first regime is due to homogenous elastic deformation. During this period, the forces and the moments in the struts monotonically increase with very little gross deformation. Note that the plastic yielding of the outer fibers of the struts already occurs at small strains (point *a*). This is due to the large bending moments that are present near the triple points. During the second regime more struts begin to yield and the plasticity spreads across the thickness and along the strut. This yielding reduces the overall stiffness of the structure compared to the elastic regime resulting in a nonlinear stress-strain response. The third regime is a very important phase of deformation and is associated with many mechanisms important in determining the foam strength. In the early part of the third regime, it can be seen from the strain maps that the strain is homogeneous throughout the sample (e.g. Fig. 2b). The structure hardens with further straining due to strain hardening of the strut material and the reorientation of the struts. Though the deformation is homogenous throughout the structure on average, in some locations the local strains develop slightly above the average strain (see Fig 2c). Initiation of the damage in the struts (at point *c*) takes place well before the peak stress is attained (at point *d* in Fig. 2). Due to this early damage, local unloading occurs and stresses get redistributed. From point *c* to *d* more struts begin to fracture at randomly distributed locations. During this period, there exist two opposing mechanisms: (i) overall hardening in some regions of the structure due to strain hardening of the strut material and strut reorientation and (ii) unloading of some regions due to the damage. The unloading regions compete with each other to maximize the overall damage rate. As a result, the damage rate in some struts decreases or even becomes zero (the first two struts from the top in the stain map *d*) with a corresponding increase in the

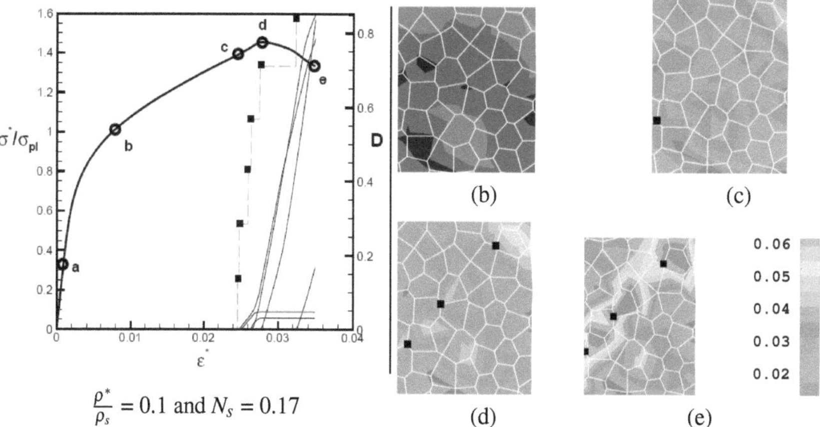

$$\frac{\rho^*}{\rho_s} = 0.1 \text{ and } N_s = 0.17$$

Fig. 2 Elasto-plastic-fracture behaviour of a two-dimensional Voronoi structure (left). The thick line is the stress-strain curve. The evolution of the damage parameter, *D*, in individual damaging struts is plotted as thin lines. The dashed line represents the total number of damaging struts with the squares representing a fracture event. The labels (*b*) to (*e*) correspond to the strain maps shown on the right. The point *a* corresponds to the first yielding event and *c* corresponds to the first damage event. The black squares in the strain maps show the location of fracture

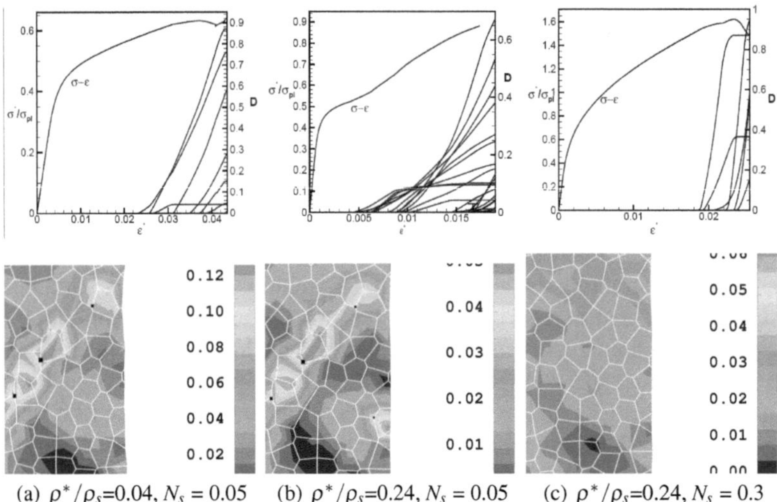

(a) ρ^*/ρ_s=0.04, N_s = 0.05   (b) ρ^*/ρ_s=0.24, N_s = 0.05   (c) ρ^*/ρ_s=0.24, N_s = 0.3

Fig. 3 Effect of relative density and strain hardening exponent on damage.The strain maps are plotted at the peak in the corresponding stress-strain curve which are immediately above each strain map

damage rate in others. This can be clearly seen from the damage evolution in the individual struts plotted in the Fig. 2. These winning regions trigger damage in the neighbouring struts. This situation appears when the random failure events localize to form a fracture path defining a future crack (see strain maps c and d in Fig. 2). At this juncture, the overall hardening rate equals the overall unloading rate due to fracture and a peak in the stress-strain curve is attained. After point d, more struts in this band satrt to fracture and all the further strain accumulates in the fracture path (see Figs. 2d and 2e). Beyond the peak unloading of the structure due to continued damage in the fracture path decreases the overall stress to zero, finally to attain a complete separation of the sample.

3.1 Influence of Strain Hardening and Relative Density on Damage

The deformation in the third regime in Fig. 2 primarily depends on the solid material's strain hardening capacity and the relative density. The effect of these two factors is presented in Fig. 3. Figs. 3a and 3b are for two different relative densities but for the same N_s=0.05. Increasing the relative density decreases the ductility (lower peak strain). Since the hardening capacity of the strut material is low, localized plastic hinges with very large curvature develop near the triple points. Due to the large curvature, the critical fracture strain will be reached soon as the thickness of the strut (relative density) is increased. This leads to an early onset of damage in an increased number of struts (see the strain map and damage curves in Fig. 3b). The increased number of damage locations now leads to two competing fracture bands. One of these early bands wins over the other; the critical struts in the dominating band continue to damage at an increased rate while those in the other band unload elastically after being partially damaged. This can be seen clearly from the damage evolution curve in the Fig. 3b where the instant of increased damage rate in some set of struts is associated with a zero damage rate in another set of struts. After this stage strain localization occurs only in this band and leads to complete unloading.

The effect of increased intrinsic material hardening for $N_s = 0.3$ and for the same relative density is shown in the Fig. 3c. With increased intrinsic solid material hardening, the localized plastic hinges disappear and a diffused plastic zone forms along the strut length. Strain hardening increases the forces in the yielded struts and in turn increasing the forces and moments in their neighbours forcing them to yield. This increased yielding results in a homogeneous strain distribution with a higher mean strain (see Fig. 3c). Since the development of high curvatures is hindered due to the extended plastic zones in the struts, the damage is postponed (giving a higher peak strain) and the total number of damage locations is also greatly reduced (compare Fig. 3b and Fig. 3c).

The left figure in Fig. 4 shows the effect of relative density on the damage for a given N_s. An earlier onset of damage can be seen with increasing relative density. The total number of damaging struts at the peak strain increases with increasing relative density. Also the rate at which damage is accumulated (the average slope of these curves) in the structure is higher at higher relative densities. However, increasing the hardening of the solid material delays the onset of damage, reduces the amount and rate of damage development (see right figure in Fig. 4). The influence of these factors on the overall response of the structure is presented in Fig. 5 in terms of the peak stress and peak strain. The peak stress scales with a power ranging between 2.09 to 1.79 (in the increasing order of N_s). The peak strain also shows strong scaling with relative density. However, it appears that averaging over larger number of realizations is required to extract the power-law scaling exponent.

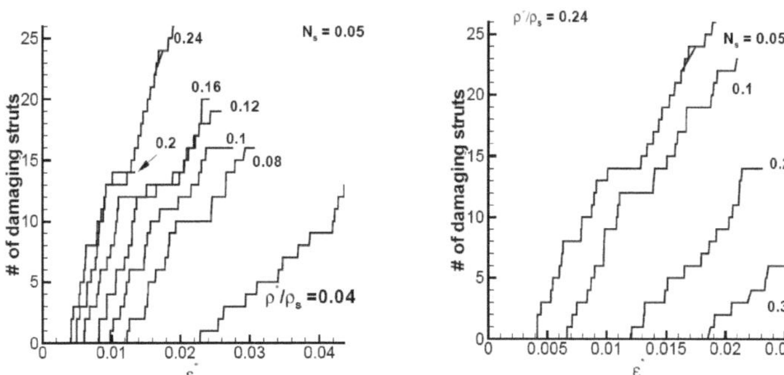

Fig. 4 Effect of relative density and N_s on the damage development

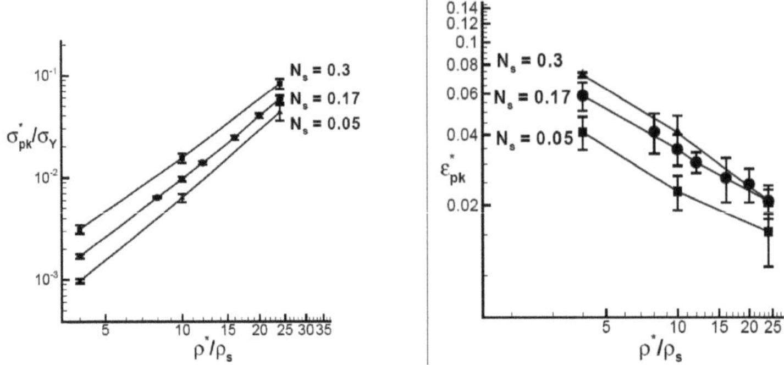

Fig. 5 Scaling of peak stress and peak strain for various N_s. The peak stress scales with relative density with a power ranging from 2.09 to 1.79 (increasing N_s)

4   Conclusions

The damage initiation and accumulation in two-dimensional Voronois is studied using a multi-scale model. The competition between various damaging elements and their role in strain localization by forming a dominant fracture band has been identified. The influence of the hardening exponent together with the relative density is is studied. Increasing the hardening capacity of the solid material results in homogeneous plastic deformation in the structure and delays the onset of damage and reduces the amount and rate at which damage develops. A strong scaling has been observed for the peak stress for various solid material hardening exponents in the range between 1.76 and 2.09.

References

1. Amsterdam, E., Onck, P.R., De Hosson, J.: Th. M., J. Mat. Sci. 40
2. Mangipudi, K.R., Onck, P.R.: Modelling Fracture of Metal Foams, Local Approach to Fracture. In: 9th European Mechanics of Materials Conference, Fontainblue, France, May 2006, p. 193 (2006)

AFCs: Active-Stress vs. Active-Strain Modeling

Paola Nardinocchi and Paolo Podio-Guidugli

Abstract. Active Fiber-reinforced Composites are artificial bodies consisting of one or more layers of parallel ceramic fibers embedded in a polymer matrix. In this paper we propose a mathematical model for their constitutive response by importing certain concepts from the modeling of fibrous living tissues and by regarding an AFC as a special piezoelectric material body whose stored energy is a weighted sum of the stored energies of fibers and matrix.

1 Introduction

Active fiber–reinforced materials may be artificial or natural. The patented acronym AFC denotes Active Fiber-reinforced Composites, that is, artificial material bodies consisting of one or more layers of parallel ceramic fibers embedded in a polymer matrix. Their main advantages over conventional piezoelectric bodies are ascribed to their toughness and flexibility, that are far greater than those of monolithic piezo-ceramics; their piezoelectric properties are also superior to those of piezopolymers. Generally, in AFCs the interactions between the electrical and mechanical material structures are linear, just as in monolithic piezoelectric materials; unlike the latter materials, electromechanical interactions occur solely along the fibers.

AFCs have been extensively studied from an experimental point view ([4], [8], [12]), but a satisfactory mathematical modeling of their response is still lacking. The main goal of this paper is to associate with any given AFC a suitable local material response, as if it were a material body and not, as it is strictly speaking, an engineered device. To do so, we import certain concepts from the modeling of

Paola Nardinocchi

Dipartimento di Ingegneria Strutturale e Geotecnica, Università di Roma "La Sapienza"
via Eudossiana 18, I-00147 Roma, Italy
e-mail: `paola.nardinocchi@uniroma1.it`

Paola Podio-Guidugli
Dipartimento di Ingegneria Civile, Università di Roma "Tor Vergata"
via del Politecnico 1, I-00133, Roma, Italy
e-mail: `ppg@uniroma2.it`

J.-F. Ganghoffer, F. Pastrone (Eds.): Mech. of Microstru. Solids 2, LNACM 50, pp. 37–48.
springerlink.com
© Springer-Verlag Berlin Heidelberg 2010

fibrous living tissues, and we regard an AFC as a special piezoelectric material body whose stored energy is a weighted sum of the stored energies of fibers and matrix.

Modeling the response of a natural AFC is delicate, because physiological issues cannot be ignored. But that is not the only difficulty: thinking of the cardiac muscle tissue to fix ideas, and making use of an engineering parlance, one must also handle both 'physical' and 'geometrical' nonlinearities, because the interactions between the electrophysiological and mechanical material structures are inherently nonlinear, and because the deformations experienced by the tissue are large (so much so that a convenient mechanical framework to set the problem appears to be finite elasticity).

For natural AFCs, two modeling formats are in use, called the *active-stress* approach and the *active-strain* approach. The former is so called after the habit, in the biological literature, to account for the ionic activation of muscle fibers by adding to the standard *passive stress* another part, called *active stress*, representing the force erogated by the muscle when it is activated [2]. In the active-strain approach (a terminology introduced in [6]; see also [10, 1]), contraction – not tension – is regarded as the direct consequence of the activation of muscle fibers; tension arises whenever contraction ($\equiv$ active strain) is hampered by some geometrical constraint.

In principle, these two modeling formats are neither equivalent nor interchangeable. In this paper (in our intentions, the first of a series), we stipulate certain equivalence conditions and investigate their consequences in the linear case, that is, the case of artificial AFCs. Our analysis, which is based on the general developments of Section 2, is summarized in Section 3. Needless to say, since in our present study physiological variables are neglected and the context is linear, our findings are not directly applicable to biological AFCs.

2   Linearly Electroelastic Materials

Linearly electroelastic materials (said differently, *piezoelectric materials*) feature linear interactions between the electrical and mechanical material structures. There are four alternative ways to choose the basic piezoelectric constitutive equations, depending on what choice of two independent variables (one mechanical, the other electrical) out of four is made [3]. The four variables are the second-order tensors $\mathbf{E}$ and $\mathbf{S}$, interpreted respectively as *strain* and *stress*, and the two vectors $\mathbf{e}$ and $\mathbf{d}$, interpreted as *electric field* and *electric displacement*. In this paper, we select the $(\mathbf{E},\mathbf{e}) \mapsto (\mathbf{S},\mathbf{d})$ and the $(\mathbf{S},\mathbf{e}) \mapsto (\mathbf{E},\mathbf{d})$ constitutive maps, because the first lends itself to the active-stress approach, the second to the active-strain approach to composite materials with piezoceramic fibers.

2.1   Active-Stress and Active-Strain Constitutive Equations

With a view toward using the active-stress approach, we introduce the invertible map

$$(\mathbf{E},\mathbf{e}) \mapsto (\mathbf{S},\mathbf{d}), \tag{1}$$

where

$$\mathbf{S} = \mathbb{C}\mathbf{E} - \mathbf{c}\mathbf{e},$$
$$\mathbf{d} = \mathbf{c}^T\mathbf{E} + \mathbf{C}\mathbf{e}. \tag{2}$$

In the active-strain approach, strain replaces stress as independent mechanical variable and a consistent switch of dependent mechanical variables is introduced. Accordingly, we consider the invertible map

$$(\mathbf{S},\mathbf{e}) \mapsto (\mathbf{E},\mathbf{d}), \tag{3}$$

where

$$\mathbf{E} = \mathbb{C}^{-1}\mathbf{S} + \mathbf{g}\mathbf{e},$$
$$\mathbf{d} = \mathbf{g}^T\mathbf{S} + \mathbf{G}\mathbf{e}. \tag{4}$$

Equations (2) prescribe that stress $\mathbf{S}$ and electric displacement $\mathbf{d}$ depend linearly on $\mathbf{E}$ and $\mathbf{e}$ through the *elasticity* and *dielectric permittivity* tensors $\mathbb{C}$ and $\mathbf{C}$ (respectively, a fourth- and a second-order tensor), and the third-order *coupling* tensor $\mathbf{c}$ and its transpose $\mathbf{c}^T$. Likewise, equations (4) prescribe that strain $\mathbf{E}$ and electric displacement $\mathbf{d}$ depend linearly on $\mathbf{S}$ and $\mathbf{e}$ through $\mathbb{C}^{-1}$ (the inverse of the elasticity tensor $\mathbb{C}$), the *coupling* tensor $\mathbf{g}$, and the alternative *dielectric permittivity* tensor $\mathbf{G}$. These constitutive tensors have symmetry and positivity properties that we postpone listing until we make our notation clear, in the next subsection.

If we rewrite the constitutive equation $(2)_1$ as follows:

$$\mathbf{S} = \mathbf{S}_p + \mathbf{S}_a, \quad \mathbf{S}_p := \mathbb{C}\mathbf{E}, \quad \mathbf{S}_a := -\mathbf{c}\mathbf{e},$$

we can regard $\mathbf{S}_a$ as the 'active' stress in a piezoelectric material, in that it is activated by turning the electric field $\mathbf{e}$ on, just as the active stress in striated muscles is turned on by the flow of calcium cations within the active-stress modeling format. Similarly, continuing to mimic what is done within the active-strain approach to muscle modeling, we may set

$$\mathbf{E}_p := \mathbb{C}^{-1}\mathbf{S}, \quad \mathbf{E}_a := \mathbf{g}\mathbf{e},$$

and read the constitutive equation $(4)_1$ as follows:

$$\mathbf{E} = \mathbf{E}_p + \mathbf{E}_a,$$

with $\mathbf{E}_a$ and $\mathbf{E}_p$ interpreted as the 'active' and 'passive' parts of the strain in a piezoelectric material.

2.2 Notational Digression

For $\mathcal{V}$ the basic three-dimensional vector space, we denote by Lin the space of all linear mappings of $\mathcal{V}$ into itself (that is to say, the space of second-order tensors), and by $\mathbb{L}\text{in}$ the space of all linear mappings of Lin into itself (that is to say, the space of fourth-order tensors). Third-order tensors such as c and g are visible in two ways, namely, as linear maps from $\mathscr{V}$ into Lin (e.g., in $(2)_1$, $\mathscr{V} \ni \mathbf{e} \mapsto \mathsf{c}\mathbf{e} \in \text{Lin}$) and as linear maps from Lin into $\mathscr{V}$ (e.g., in $(4)_2$, $\text{Lin} \ni \mathbf{S} \mapsto \mathsf{g}^T\mathbf{S} \in \mathscr{V}$).

Let $\{\mathbf{c}_i, i = 1,2,3\}$ be a left-oriented triad of orthonormal vectors, a cartesian basis for $\mathscr{V}$, and let $\{\mathbf{c}_i \otimes \mathbf{c}_j, i, j = 1,2,3\}$ be the corresponding basis for Lin. The components of a third-order tensor – say, the cartesian components of the coupling tensor c – are defined as follows:

$$\mathsf{c} = c_{ijk}\,\mathbf{c}_i \otimes \mathbf{c}_j \otimes \mathbf{c}_k;$$

moreover,

$$\mathsf{c}^T = c_{ijk}\,\mathbf{c}_k \otimes \mathbf{c}_i \otimes \mathbf{c}_j \quad (\text{or rather, } (\mathsf{c}^T)_{kij} = (\mathsf{c})_{ijk} = c_{ijk}).$$

Thus, the linear action of a third-order tensor on, respectively, a vector and a second-order tensor are defined as follows:

$$\mathsf{c}\mathbf{e} = (\mathbf{c}_k \cdot \mathbf{e})\,c_{ijk}\mathbf{c}_i \otimes \mathbf{c}_j = (c_{ijk}e_k)\,\mathbf{c}_i \otimes \mathbf{c}_j$$

and

$$\mathsf{c}^T\mathbf{E} = (\mathbf{c}_i \otimes \mathbf{c}_j \cdot \mathbf{E})\,c_{ijk}\mathbf{c}_k = (c_{ijk}E_{ij})\,\mathbf{c}_k;$$

the second-order tensor $\mathsf{c}\mathbf{e}$ and the vector $\mathsf{c}^T\mathbf{E}$ are such that

$$\mathsf{c}\mathbf{e} \cdot \mathbf{E} = \mathbf{e} \cdot \mathsf{c}^T\mathbf{E}.\,^{[1]}$$

Likewise, the linear action of a fourth-order tensor

$$\mathbb{C} = C_{ijkl}\,\mathbf{c}_i \otimes \mathbf{c}_j \otimes \mathbf{c}_k \otimes \mathbf{c}_l$$

on a second-order tensor

$$\mathbf{E} = E_{ij}\,\mathbf{c}_i \otimes \mathbf{c}_j$$

is defined by

$$\mathbb{C}\mathbf{E} = (\mathbf{c}_k \otimes \mathbf{c}_l \cdot \mathbf{E})\,C_{ijkl}\,\mathbf{c}_i \otimes \mathbf{c}_j = (C_{ijkl}E_{kl})\,\mathbf{c}_i \otimes \mathbf{c}_j.$$

Moreover, the linear action of a fourth-order tensor $\mathbb{A}$ on a third-order tensor a is defined by

$$\begin{aligned}
\mathbb{A}\mathsf{a} &= (A_{lmpq}\mathbf{c}_l \otimes \mathbf{c}_m \otimes \mathbf{c}_p \otimes \mathbf{c}_q)(a_{ijk}\,\mathbf{c}_i \otimes \mathbf{c}_j \otimes \mathbf{c}_k) \\
&= ((\mathbf{c}_p \cdot \mathbf{c}_i)(\mathbf{c}_q \cdot \mathbf{c}_j)A_{lmpq}a_{ijk})\,\mathbf{c}_l \otimes \mathbf{c}_m \otimes \mathbf{c}_k \\
&= (A_{lmij}a_{ijk})\,\mathbf{c}_l \otimes \mathbf{c}_m \otimes \mathbf{c}_k.
\end{aligned}$$

---

[1] With slight abuse of notation, the centered dot on the left side denotes inner product of second-order tensors, that on the left inner product of vectors.

As to composition products, for tensors of even order they are defined as customary:

$$
\begin{aligned}
\mathbf{AB} &= (A_{ij}\mathbf{c}_i \otimes \mathbf{c}_j)(B_{hk}\mathbf{c}_h \otimes \mathbf{c}_k) \\
&= \big((\mathbf{c}_j \cdot \mathbf{c}_h)A_{ij}B_{hk}\big)\,\mathbf{c}_i \otimes \mathbf{c}_k \\
&= (A_{ih}B_{hk})\,\mathbf{c}_i \otimes \mathbf{c}_k
\end{aligned}
$$

and

$$
\begin{aligned}
\mathbb{AB} &= (A_{ijhk}\,\mathbf{c}_i \otimes \mathbf{c}_j \otimes \mathbf{c}_h \otimes \mathbf{c}_k)(B_{lmpq}\,\mathbf{c}_l \otimes \mathbf{c}_m \otimes \mathbf{c}_p \otimes \mathbf{c}_q) \\
&= \big((\mathbf{c}_k \cdot \mathbf{c}_l)(\mathbf{c}_h \cdot \mathbf{c}_m)A_{ijhk}B_{lmpq}\big)\,\mathbf{c}_i \otimes \mathbf{c}_j \otimes \mathbf{c}_p \otimes \mathbf{c}_q \\
&= (A_{ijhk}B_{khpq})\,\mathbf{c}_i \otimes \mathbf{c}_j \otimes \mathbf{c}_p \otimes \mathbf{c}_q\,.
\end{aligned}
$$

The less common composition products of second-, third- and fourth-order tensors are denoted by the symbol $\circ$, and defined as follows:

$$
\begin{aligned}
\mathbf{a} \circ \mathbf{A} &= (a_{ijk}\,\mathbf{c}_i \otimes \mathbf{c}_j \otimes \mathbf{c}_k)(A_{lm}\,\mathbf{c}_l \otimes \mathbf{c}_m) \\
&= \big((\mathbf{c}_k \cdot \mathbf{c}_l)\,a_{ijk}A_{lm}\big)\,\mathbf{c}_i \otimes \mathbf{c}_j \otimes \mathbf{c}_m \\
&= (a_{ijk}b_{km})\,\mathbf{c}_i \otimes \mathbf{c}_j \otimes \mathbf{c}_m\,;
\end{aligned}
$$

$$
\begin{aligned}
\mathbf{A} \circ \mathbf{a} &= (a_{lm}\,\mathbf{c}_l \otimes \mathbf{c}_m)(a_{ijk}\,\mathbf{c}_i \otimes \mathbf{c}_j \otimes \mathbf{c}_k) \\
&= \big((\mathbf{c}_m \cdot \mathbf{c}_i)\,a_{ijk}A_{lm}\big)\,\mathbf{c}_l \otimes \mathbf{c}_j \otimes \mathbf{c}_k \\
&= (A_{li}\,a_{ijk})\,\mathbf{c}_l \otimes \mathbf{c}_j \otimes \mathbf{c}_k\,;
\end{aligned}
$$

$$
\begin{aligned}
\mathbf{a} \circ \mathbf{b} &= (a_{ijk}\,\mathbf{c}_i \otimes \mathbf{c}_j \otimes \mathbf{c}_k)(b_{lmp}\,\mathbf{c}_l \otimes \mathbf{c}_m \otimes \mathbf{c}_p) \\
&= \big((\mathbf{c}_k \cdot \mathbf{c}_l)(\mathbf{c}_j \cdot \mathbf{c}_m)\,a_{ijk}b_{lmp}\big)\,\mathbf{c}_i \otimes \mathbf{c}_p \\
&= (a_{ijk}b_{kjp})\,\mathbf{c}_i \otimes \mathbf{c}_p\,;
\end{aligned}
$$

$$
\begin{aligned}
\mathbf{a} \circ \mathbb{A} &= (a_{ijk}\,\mathbf{c}_i \otimes \mathbf{c}_j \otimes \mathbf{c}_k)(\mathbb{A}_{lmpq}\mathbf{c}_l \otimes \mathbf{c}_m \otimes \mathbf{c}_p \otimes \mathbf{c}_q) \\
&= \big((\mathbf{c}_k \cdot \mathbf{c}_l)(\mathbf{c}_j \cdot \mathbf{c}_m)\,a_{ijk}A_{lmpq}\big)\,\mathbf{c}_i \otimes \mathbf{c}_p \otimes \mathbf{c}_q \\
&= (a_{ijk}A_{kjpq})\,\mathbf{c}_i \otimes \mathbf{c}_p \otimes \mathbf{c}_q
\end{aligned}
$$

2.3   *Symmetry and Positivity of the Constitutive Tensors*

Hereafter we collect the component versions of the symmetry and positivity properties of the constitutive tensors entering the response laws (2) and (4):

- *elasticity tensor* $\mathbb{C}$:

$$
C_{ijhk} = C_{jihk} = C_{ijkh}, \quad C_{ijhk} = C_{hkij}, \quad \mathbf{A} \cdot \mathbb{C}\mathbf{A} > 0 \quad \forall \mathbf{A} \in \mathrm{Lin} \setminus \{\mathbf{0}\};
$$

- *coupling tensors* $\mathbf{c}, \mathbf{g}$:

$$
c_{ijk} = c_{jik}, \quad g_{ijk} = g_{jik}\,;
$$

- *permittivity tensors* $\mathbf{C}, \mathbf{G}$:

$$
\begin{aligned}
C_{ij} &= C_{ji}, \quad \mathbf{a} \cdot \mathbf{C}\mathbf{a} > 0 \quad \forall \mathbf{a} \in \mathscr{V} \setminus \{\mathbf{0}\}, \\
G_{ij} &= G_{ji}, \quad \mathbf{a} \cdot \mathbf{G}\mathbf{a} > 0 \quad \forall \mathbf{a} \in \mathscr{V} \setminus \{\mathbf{0}\}.
\end{aligned}
$$

2.4 *Equivalent Constitutive Equations*

One may ask: when is that two lists $\{\mathbb{C}, \mathsf{c}, \mathbf{C}\}$ and $\{\mathbb{C}^{-1}, \mathsf{g}, \mathbf{G}\}$ of constitutive tensors characterize the same piezoelectric material? We propose to regard the constitutive equations of type (2) and (4) constructed by the use of those two lists as *equivalent* if the following two conditions hold:

for all symmetric tensors $\mathbf{E}$ and for all vectors $\mathbf{e}$,
(**i**) *both* $(2)_1$ *and* $(4)_1$ *deliver the same stress*, that is to say,

$$\mathbb{C}\mathbf{E} - \mathsf{c}\mathbf{e} = \mathbf{S} = \mathbb{C}(\mathbf{E} - \mathsf{g}\mathbf{e}); \tag{5}$$

(**ii**) *both* $(2)_2$ *and* $(4)_2$ *deliver the same electric displacement*, that is to say, if

$$\mathsf{c}^T\mathbf{E} + \mathbf{C}\mathbf{e} = \mathbf{d} = \mathsf{g}^T(\mathbb{C}(\mathbf{E} - \mathsf{g}\mathbf{e})) + \mathbf{G}\mathbf{e}. \tag{6}$$

Condition (5) is verified if

$$\mathsf{c}\mathbf{e} = \mathbb{C}(\mathsf{g}\mathbf{e}) \quad \text{for all vectors } \mathbf{e},$$

i.e., due to some of the symmetry properties stipulated for $\mathbb{C}$ and g, if

$$\mathsf{c} = \mathbb{C} \circ \mathsf{g}. \tag{7}$$

With this,
$$\mathsf{c}^T\mathbf{E} = \mathsf{g}^T(\mathbb{C}\mathbf{E}) \quad \text{for all symmetric tensors } \mathbf{E};$$

hence, condition (6) is verified if

$$\mathbf{C}\mathbf{e} = \mathbf{G}\mathbf{e} - \mathsf{g}^T(\mathbb{C}(\mathsf{g}\mathbf{e})) \quad \text{for all vectors } \mathbf{e},$$

i.e., if
$$\mathbf{C} = \mathbf{G} - \mathsf{g}^T \circ \mathbb{C} \circ \mathsf{g}. \tag{8}$$

REMARK. Note the following consequence of the equivalence condition (7):

$$\mathbf{S}_a = -\mathbb{C}\mathbf{E}_a. \tag{9}$$

Thus, when the active-stress and active-strain approaches to the constitutive modeling of piezoelectric materials are made equivalent, there is a one-to-one correspondence between the active parts of strain and stress.

2.5 *Boundary Working, Internal Power, and Stored Energy*

When thermal phenomena are ignored, the state of a piezoelectric body can be described in terms of two fields, the *displacement* $\mathbf{u}$ and the *electrical potential* ϕ, respectively, a vector and a scalar field over the region $\mathscr{B}$ the body occupies; the corresponding strain and electrical fields $\mathbf{E}$ and $\mathbf{e}$ are:

$$\mathbf{E} = \operatorname{sym}\nabla\mathbf{u}, \quad \mathbf{e} = -\nabla\phi \tag{10}$$

(here sym denotes the operation of taking the symmetric part of a second-order tensor). We read in [11], Chapter 5, that, for a body part $\mathscr{P}$ with boundary $\partial\mathscr{P}$ of outer normal $\mathbf{n}$, the *boundary working* is "rate at which work is done by the surface tractions acting across $\partial\mathscr{P}$ less the flux of electric energy outward across $\partial\mathscr{P}$", namely,

$$\int_{\partial\mathscr{P}} \left(\mathbf{Sn}\cdot\dot{\mathbf{u}} - \phi(\dot{\mathbf{d}}\cdot\mathbf{n})\right).$$

Consequently, given that the electric displacement is taken to be divergenceless in the present context, the *internal power* expended over a test pair $(\delta\mathbf{u}, \delta\mathbf{d})$ is defined to be:

$$P(\mathscr{P})[\delta\mathbf{u}, \delta\mathbf{d}] := \int_{\mathscr{P}} \left(\mathbf{S}\cdot\operatorname{sym}\nabla(\delta\mathbf{u}) + \mathbf{e}\cdot\delta\mathbf{d}\right).$$

The accompanying *dissipation principle* states that the time rate of the *stored energy* density per unit volume w satisfies:

$$\dot{w} \leq \mathbf{S}\cdot\dot{\mathbf{E}} + \mathbf{e}\cdot\dot{\mathbf{d}},$$

for all processes and whatever their local continuation; equivalently, in terms of the *electric enthalphy h*, that

$$\dot{h} \leq \mathbf{S}\cdot\dot{\mathbf{E}} - \mathbf{d}\cdot\dot{\mathbf{e}}, \quad h := w - \mathbf{d}\cdot\mathbf{e}. \tag{11}$$

We define the *stored energy* density per unit volume as follows:

$$w = \frac{1}{2}\mathbf{S}\cdot\mathbf{E} + \frac{1}{2}\mathbf{d}\cdot\mathbf{e}. \tag{12}$$

An use of the constitutive equations (4) yields the stored energy mapping appropriate to an active-stess approach to modeling piezoelectric materials, namely,

$$(\mathbf{E}, \mathbf{e}) \mapsto w = \widehat{w}(\mathbf{E}, \mathbf{e}) := \frac{1}{2}\mathbb{C}\mathbf{E}\cdot\mathbf{E} + \frac{1}{2}\mathbf{Ce}\cdot\mathbf{e}; \tag{13}$$

needless to say, positivity of $\mathbb{C}$ and $\mathbf{C}$ is a necessary and sufficient condition for $\widehat{w}$ to be positive-valued. Note that the corresponding enthalphy mapping is:

$$(\mathbf{E}, \mathbf{e}) \mapsto h = \widehat{h}(\mathbf{E}, \mathbf{e}) := \frac{1}{2}\mathbb{C}\mathbf{E}\cdot\mathbf{E} - \frac{1}{2}\mathbf{Ce}\cdot\mathbf{e} - \mathbf{E}\cdot\mathsf{ce}. \tag{14}$$

Were we to start with this definition together with the dissipation principle (11), then we would readily obtain:

$$\partial_{\mathbf{E}}\widehat{h}(\mathbf{E}, \mathbf{e}) = \mathbb{C}\mathbf{E} - \mathsf{ce} = \mathbf{S},$$
$$-\partial_{\mathbf{e}}\widehat{h}(\mathbf{E}, \mathbf{e}) = \mathsf{c}^T\mathbf{E} + \mathbf{Ce} = \mathbf{d}.$$

Interestingly, for c and $\mathbf{C}$ compliant with the equivalence conditions (7) and (8), we find that, alternatively, the stored energy density is also delivered by the following mapping:

$$(\mathbf{S},\mathbf{e}) \;\mapsto\; w = \check{w}(\mathbf{S},\mathbf{e}) := \frac{1}{2}\mathbf{S}\cdot\mathbb{C}^{-1}\mathbf{S} + \frac{1}{2}\mathbf{G}\mathbf{e}\cdot\mathbf{e} + \mathbf{S}\cdot\mathsf{g}\mathbf{e}. \tag{15}$$

It is easy to check that, within an active-stain approach, (12) and (4) yield just (15).

2.6   *Material Symmetry*

An orthogonal tensor $\mathbf{Q}$ is a *material symmetry transformation* for a piezoelectric material whose response from a fixed reference placement is described by the elasticity, coupling, and dielectric tensors $\mathbb{C}$, c, and $\mathbf{C}$, if it so happens that

$$\begin{aligned}
\mathbf{Q}(\mathbb{C}\mathbf{E} - \mathsf{c}\mathbf{e})\mathbf{Q}^{T} &= \mathbb{C}\mathbf{Q}\mathbf{E}\mathbf{Q}^{T} - \mathsf{c}\mathbf{Q}\mathbf{e}, \\
\mathbf{Q}(\mathsf{c}^{T}\mathbf{E} + \mathbf{C}\mathbf{e}) &= \mathsf{c}^{T}(\mathbf{Q}\mathbf{E}\mathbf{Q}^{T}) + \mathbf{C}(\mathbf{Q}\mathbf{e}),
\end{aligned} \tag{16}$$

for all symmetric tensors $\mathbf{E}$ and for all vectors $\mathbf{e}$; as is well known, the collection of all such orthogonal tensors has the group structure. To give (16) a more compact form, we introduce the orthogonal conjugation operator $\mathbb{Q}$ associated to the orthogonal tensor $\mathbf{Q}$:

$$\mathbb{Q}\mathbf{A} := \mathbf{Q}\mathbf{A}\mathbf{Q}^{T}$$

for all second-order tensors $\mathbf{A}$. Then, condition $(16)_1$ can be written as:

$$\mathbb{Q}\mathbb{C} = \mathbb{C}\mathbb{Q} \quad \text{and} \quad \mathbb{Q}\circ\mathsf{c} = \mathsf{c}\circ\mathbf{Q}; \tag{17}$$

likewise, condition $(16)_2$ reads:

$$\mathbf{Q}\mathbf{C} = \mathbf{C}\mathbf{Q} \quad \text{and} \quad \mathbf{Q}\circ\mathsf{c}^{T} = \mathsf{c}^{T}\circ\mathbb{Q}. \tag{18}$$

The *material symmetry group* of a piezoelectric material whose response is specified by the list $\{\mathbb{C},\mathsf{c},\mathbf{C}\}$ of constitutive tensors is the collection $\mathscr{G}$ of orthogonal tensors such as to satisfy both (17) and (18).

REMARK. In case the response of a piezoelectric material is expressed in terms of a list $\{\mathbb{C}^{-1},\mathsf{g},\mathbf{G}\}$ by means of the constitutive equations (4), an orthogonal tensor $\mathbf{Q}$ is a material symmetry transformation if the coupling tensor g and the dielectric tensor $\mathbf{G}$ satisfy the same conditions as, respectively, c and $\mathbf{C}$ (that is to say, g satisfies $(17)_2$ and $(18)_2$, and $\mathbf{G}$ satisfies $(18)_1$). As expected, two lists being equivalent in the sense of Subsection 2.5 share the same material symmetry group.

2.7   *Transversely Isotropic Piezoelectric Materials*

Whenever a material symmetry group $\mathscr{G}$ is specified, quantification of relations (17) and (18) over all elements of $\mathscr{G}$ yields $\mathscr{G}$−specific representation formulae for the constitutive tensors $\mathbb{C}$, c, and $\mathbf{C}$. An instance important to our present purposes is when $\mathscr{G}$ is the continuous group of all rotations about the axis spanned by a fixed unit vector ($\mathbf{c}_3$, say), identifying the class of piezoelectric materials that are *transversely isotropic* with respect to that axis. We take from [7], Section 16, the following representation formula for the elasticity tensor:

$$\begin{aligned}
\mathbb{C} &= \gamma_1(\mathbf{C}_1 \otimes \mathbf{C}_1 + \mathbf{C}_2 \otimes \mathbf{C}_2) + \gamma_2(\mathbf{C}_3 \otimes \mathbf{C}_3 + \mathbf{C}_4 \otimes \mathbf{C}_4) \\
&= \delta_1 \mathbf{P} \otimes \mathbf{P} + \delta_2(\mathbf{P} \otimes \mathbf{P}^\perp + \mathbf{P}^\perp \otimes \mathbf{P}) + \delta_3 \mathbf{P}^\perp \otimes \mathbf{P}^\perp ;
\end{aligned} \tag{19}$$

where

$$\begin{aligned}
\sqrt{2}\,\mathbf{C}_\alpha &= \mathbf{c}_\alpha \otimes \mathbf{c}_3 + \mathbf{c}_3 \otimes \mathbf{c}_\alpha , \\
\sqrt{2}\,\mathbf{C}_3 &= \mathbf{c}_1 \otimes \mathbf{c}_2 + \mathbf{c}_2 \otimes \mathbf{c}_1 , \\
\sqrt{2}\,\mathbf{C}_4 &= \mathbf{c}_1 \otimes \mathbf{c}_1 - \mathbf{c}_2 \otimes \mathbf{c}_2 , \\
\mathbf{P} &= \mathbf{c}_3 \otimes \mathbf{c}_3 , \\
\sqrt{2}\,\mathbf{P}^\perp &= \mathbf{I} - \mathbf{P}
\end{aligned} \tag{20}$$

(here, $\alpha = 1,2$; note that the six tensors $\{\mathbf{C}_1,\ldots,\mathbf{C}_4,\mathbf{P},\mathbf{P}^\perp\}$ form an orthornormal basis for the subspace of Lin of all symmetric tensors). As to the dielectric tensors, it is not difficult to show that they can be represented as follows:

$$\begin{aligned}
\mathbf{C} &= \gamma \mathbf{P} + \bar{\gamma} \mathbf{P}^\perp , \\
\mathbf{G} &= \delta \mathbf{P} + \bar{\delta} \mathbf{P}^\perp .
\end{aligned} \tag{21}$$

With a bit more work, representations for the coupling tensors are arrived at; they are:

$$\begin{aligned}
\mathsf{c} &= \alpha_1\,\mathbf{C}_\alpha \otimes \mathbf{c}_\alpha + \alpha_2\,\mathbf{P}^\perp \otimes \mathbf{c}_3 + \alpha_3\,\mathbf{P} \otimes \mathbf{c}_3 , \\
\mathsf{g} &= \beta_1\,\mathbf{C}_\alpha \otimes \mathbf{c}_\alpha + \beta_2\,\mathbf{P}^\perp \otimes \mathbf{c}_3 + \beta_3\,\mathbf{P} \otimes \mathbf{c}_3
\end{aligned} \tag{22}$$

(here, summation over the index $\alpha = 1,2$ is to be understood). The algebraic symmetries listed in Subsection 2.2 are easily checked. As to positivity, $\mathbb{C}$ is positive iff γ_1, γ_2, and δ_1, are positive, and, in addition,

$$\delta_1 \delta_3 - {\delta_2}^2 > 0; \tag{23}$$

$\mathbf{C}$ is positive iff γ and $\bar{\gamma}$ are positive; and similarly for $\mathbf{G}$.

For a transversely isotropic piezoelectric material, the equivalence representation formulae (7) and (8) give:

$$\begin{aligned}
\mathsf{c} &= \alpha_1\,\mathbf{C}_\alpha \otimes \mathbf{c}_\alpha + \alpha_2\,\mathbf{P}^\perp \otimes \mathbf{c}_3 + \alpha_3\,\mathbf{P} \otimes \mathbf{c}_3 \\
&= \beta_1 \gamma_1\,\mathbf{C}_\alpha \otimes \mathbf{c}_\alpha + (\beta_2 \delta_3 + \beta_3 \delta_2)\,\mathbf{P}^\perp \otimes \mathbf{c}_3 + (\beta_2 \delta_2 + \beta_3 \delta_1)\,\mathbf{P} \otimes \mathbf{c}_3
\end{aligned} \tag{24}$$

and

$$\mathbf{C} = \gamma\mathbf{P} + \bar{\gamma}\mathbf{P}^{\perp}$$
$$= \left(\delta - (\beta_2{}^2\delta_3 + 2\beta_2\beta_3\delta_2 + \beta_3{}^2\delta_1)\right)\mathbf{P} + \left(\bar{\delta} - \beta_1{}^2\gamma_1\right)\mathbf{P}^{\perp}. \tag{25}$$

Thus, for $\mathbf{C}$ to be positive, the material moduli must be such that

$$\delta - (\beta_2{}^2\delta_3 + 2\beta_2\beta_3\delta_2 + \beta_3{}^2\delta_1) > 0, \quad \bar{\delta} - \beta_1{}^2\gamma_1 > 0; \tag{26}$$

interestingly, these two conditions involve all material moduli except the shear modulus γ_2.

3   AFCs as Piezoelectric Materials

To model the electroelastic response of an active fiber-reinforced composite, we assume that:
(i) only fibers are capable of electromechanical interactions;
(ii) the composite's effective stored energy is a weighted sum of the stored energies of matrix and fibers:

$$w_c = \varphi w_m + (1 - \varphi)w_f, \quad \varphi \in (0, 1) \tag{27}$$

(see [9], [5], where w_m and w_f are weighted the same). Needless to say, with this recipe for w_c, the composite's material symmetry group will be the smaller between $\mathcal{G}_m$ and $\mathcal{G}_f$.

As to the matrix, it is usually safe to take it to be an isotropic material:

$$w_m(\mathbf{S}) = \frac{1}{2}\mathbf{S} \cdot \mathbb{C}_m^{-1}\mathbf{S}, \quad \mathbb{C}_m = 2\mu\,\mathbb{I} + \lambda\mathbf{I} \otimes \mathbf{I} \tag{28}$$

with $\mathbb{I}$ the identity in $\mathbb{L}\text{in}$; more general response functions can be accommodated without problems.

As to fibers, we assign them a transversely isotropic response that is special under a number of respects:
(iii) the transverse isotropy axis is aligned with the fiber direction, which we choose to be $\mathbf{c}_3$;
(iv) the elastic modulus δ_2 is null;
(v) the coupling moduli β_1 and β_2 are null.

Thus, in view of (19)-(20), we have that

$$\mathbb{C}_f = \gamma_1(\mathbf{C}_1 \otimes \mathbf{C}_1 + \mathbf{C}_2 \otimes \mathbf{C}_2) + \gamma_2(\mathbf{C}_3 \otimes \mathbf{C}_3 + \mathbf{C}_4 \otimes \mathbf{C}_4) + \delta_1\mathbf{P} \otimes \mathbf{P} + \delta_3\mathbf{P}^{\perp} \otimes \mathbf{P}^{\perp}; \tag{29}$$

in view of, respectively, $(22)_2$ and $(24)_2$, that

$$\mathbf{g}_f = \beta_3\mathbf{P} \otimes \mathbf{c}_3, \quad \mathbf{c}_f = \mathbb{C}_f \circ \mathbf{g}_f = \beta_3\delta_1\mathbf{P} \otimes \mathbf{c}_3; \tag{30}$$

and, finally, in view of $(21)_2$ and $(25)_2$, that

$$\mathbf{C}_f = \mathbf{G}_f - \beta_3{}^2\delta_1\mathbf{P}, \quad \mathbf{G}_f = \delta\mathbf{P} + \bar{\delta}\mathbf{P}^{\perp}. \tag{31}$$

To sum up, four *elasticities* (γ_1, γ_2, δ_1, and δ_3), two *permittivities* (δ and $\bar{\delta}$), and one *coupling modulus* (β_3) determine the fiber response; two more elasticities (λ and μ) are needed to specify the matrix response; one further parameter (ϕ) fixes the matrix volume-fraction. The active part of the strain depends only on the coupling modulus β_3, while the active stress depends also on the elasticity δ_1:

$$\mathbf{E}_a = \beta_3(\mathbf{e}\cdot\mathbf{c}_3)\mathbf{P}, \quad \mathbf{S}_a = -\delta_1\mathbf{E}_a. \tag{32}$$

Component-wise, relations (2) read, respectively,

$$\begin{aligned}
S_{11} &= \tfrac{1}{2}(\gamma_2 + \delta_3)E_{11} + \tfrac{1}{2}(-\gamma_2 + \delta_3)E_{22}, \\
S_{22} &= \tfrac{1}{2}(-\gamma_2 + \delta_3)E_{11} + \tfrac{1}{2}(\gamma_2 + \delta_3)E_{22}, \\
S_{12} &= \gamma_2 E_{12}, \quad S_{\alpha 3} = \gamma_1 E_{\alpha 3} \ (\alpha = 1,2), \\
S_{33} &= \delta_1(E_{33} - \beta_3 e_3),
\end{aligned} \tag{33}$$

and

$$\begin{aligned}
d_\alpha &= \bar{\delta}e_\alpha \ (\alpha = 1,2), \\
d_3 &= \left((\delta - \beta_3{}^2\delta_1)e_3 + \delta_1\beta E_{33}\right).
\end{aligned} \tag{34}$$

REMARKS. 1.Typically, in applications involving AFCs, the electric field has everywhere the direction of fibers ($e_\alpha \equiv 0$), so that knowledge of the permittivity $\bar{\delta}$ is immaterial. Moreover – and tacitly! – the elasticity γ_1 is taken null, a measure that we interpret as a brute-force way to guarantee that the angle between the transverse-isotropy direction $\mathbf{c}_3$ and all material directions orthogonal to it is preserved, whatever the deformational vicissitudes. As a result of these two simplifications, the fiber response is specified in terms of only five material moduli – three elasticities, one permittivity and one coupling modulus – whose values one can usually pick up from the experimental literature.

2. At matrix/fiber boundaries, certain continuity conditions must be imposed, typically to guarantee that no slip occurs. Such conditions may be written in a form that amounts to a condition of constitutive consistency between fibers and matrix: e.g., for $\mathbf{n}$ a unit vector perpendicular to the fiber direction, one is led to insist that

$$\mathbf{S}_f(\mathbf{E},\mathbf{e})\mathbf{n} = \mathbf{S}_m(\mathbf{E},\mathbf{e})\mathbf{n}$$

for all pairs $(\mathbf{E},\mathbf{e})$.

References

1. Cherubini, C., Filippi, S., Nardinocchi, P., Teresi, L.: An Electromechanical Model of Cardiac Tissue: Constitutive Issues and Electrophysiological Effects. Progr. Biophys. Mol. Biol. (2008) (to appear)

2. Hunter, P.J., McCulloch, A.D., ter Keurs, H.E.D.J.: Modelling the mechanical properties of cardiac muscle. Progr. Biophys. Mol. Biol. 69, 289–331 (1998)
3. Ikeda, T.: Fundamentals of Piezoelectricity. Oxford University Press, New York (1990)
4. Melnykowyczl, M., Kornmann, X., Huber, C., Barbezat, M., Brunner, A.J.: Performances of integrated active fiber composites in fiber reinforced epoxy laminates. Smart Mater. Struct. 15, 204–212 (2006)
5. Merodio, J., Ogden, R.W.: Mechanical response of fiber-reinforced incompressible non-linearly elastic solids. Int. J. Non-Lin. Mech. 40, 213–227 (2005)
6. Nardinocchi, P., Teresi, L.: On the active response of soft living tissues. J. Elasticity 88, 27–39 (2007)
7. Podio-Guidugli, P.: A Primer in Elasticity. Kluwer, Dordrecht (2000)
8. Rossetti, G.A., Pizzochero, A., Bent, A.A.: Recent Advances in Active Fiber Composite Technology, pp. 753–756. IEEE, Los Alamitos (2001)
9. Spencer, A.J.M.: Continuum Theory of the Mechanics of Fibre-Reinforced Composites. Springer, Wien (1984)
10. Stålhand, J., Klarbring, A., Holzapfel, G.A.: Smooth muscle contraction: Mechanochemical formulation for homogeneous finite strains. Progr. Biophys. Mol. Biol. 96, 465–481 (2008)
11. Tiersten, H.F.: Linear Piezoelectric Plate Vibrations. Plenum Press, New York (1969)
12. Wickramasinghe, V.K., Hagood, N.W.: Material characterization of active fiber composites for integral twist-actuated rotor blade application. Smart Mater. Struct. 13, 1155–1165 (2004)

On Eshelby Tensors, Thermodynamics and Calculus of Variations

Jean-François Ganghoffer

Abstract. The connections between the notion of Eshelby tensor and the variation of Hamiltonian like action integrals are investigated, in connection with the thermodynamics of continuous open bodies exchanging mass, heat and work with their surrounding. Considering first a homogeneous representative volume element (RVE), it is shown that a possible choice of the Lagrangian density is the material derivative of a suitable thermodynamic potential. The Euler equations of the so built action integral are the state laws written in rate form. As the consequence of the optimality conditions of the resulting Jacobi action, the vanishing of the surface contribution resulting from the general variation of this Hamiltonian action leads to the well-known Gibbs-Duhem condition. A general three-field variational principle describing the equilibrium of heterogeneous systems is next written, based on the zero potential, the stationnarity of which delivers a balance law for a generalized Eshelby tensor in a thermodynamic context. Adopting the rate of the grand potential as the lagrangian density, a generalized Gibbs-Duhem condition is obtained as the transversality condition of the thermodynamic action integral, considering a solid body with a movable boundary.

Keywords: thermodynamics of open systems; transversality conditions; Eshelby tensors; Noether's theorem; Gibbs-Duhem condition.

1 Introduction

In a celebrated article, Eshelby (1951) introduced the concept of the energy-momentum tensor, which has been involved since then in several theories such as Eshelbian mechanics and the related notion of configurational forces, including not only continuum mechanics, but also electromagnetoelasticity in the large, (Maugin, 1995). Fundamental works along these line of thoughts include those of Maugin (1993, 1995), Kienzler and Herrmann (2000), Gurtin (2000), and applications of these concepts to the analysis of different of material inhomogeneities have been done in the recent years, see the review article in Gross et al. (2003).

Jean-François Ganghoffer
LEMTA – ENSEM
2, Avenue de la Forêt de Haye
BP 160 54504 Vandoeuvre CEDEX France
e-mail: `jean-francois.Ganghoffer@ensem.inpl-nancy.fr`

J.-F. Ganghoffer, F. Pastrone (Eds.): Mech. of Microstru. Solids 2, LNACM 50, pp. 49–62.
springerlink.com   © Springer-Verlag Berlin Heidelberg 2010

The derivation of the configurational force balance can be obtained following two main alternative routes: either from a pull-back of the classical balance laws on the material manifold (Maugin, 1993), or using the notion of translational invariance, in articulation with Noether's theorem (1918), as exemplified in (Kienzler and Hermann, 2000). Configurational mechanics (otherwise coined mechanics of material forces) has witnessed a revival in the last decade, and it has been a very active field of research in the recent period: in addition to previous references, one can mention the works of (Kuhl and Steinmann, 2004) focusing on material forces for open systems, (Lubarda and Markenscoff, 2007) related to dual conservation laws in micropolar elasticity, evaluation of configurational forces in multiplicative elastoplasticity (Menzel and Steinmann, 2007), the consideration of material forces in dynamic fracture (Fagerström and Larsson, 2008) or (Agiasofitou and Kalpakides, 2006), and furthermore the work of Steinmann (2008) highlighting the role of boundary potential energies.

In the present contribution, the connections between the Eshelby tensor and the variation of Hamiltonian like action integrals will be investigated, considering the general framework of the thermodynamics of continuous open bodies exchanging mass, heat and work with their surrounding. Such a framework clearly has interesting applications in the field of biomechanics. The construction of Eshelby tensors from suitable thermodynamic extremum principles shall be first evidenced, in articulation with the concepts and methods of the calculus of variations. It is here worthwhile mentioning the clarifying early work of Edelen (1981), who examined the role of transversality conditions on movable internal and external boundaries. In a more recent period, Honein et al. (1991) established a method to obtain conservation laws for dissipative phenomena. Application of this method to dissipative thermoelasticity was done by (Chien and Herrmann, 1996). In the same line of thoughts, conservation laws involving Eshelby stress in the case of the dissipationless theory of thermoelasticity have been written in (Kalpakides and Maugin, 2004).

The convention of summation of any repeated index in a monomial is systematically used in the sequel. Vectors and tensors are represented as boldface symbols. The symbol $\mathbf{I}$ denotes the second order identity tensor. The transpose of any tensor $\mathbf{A}$ is the tensor noted $\mathbf{A}^{\mathrm{T}}$. The double inner product of two second order tensors $\mathbf{A}, \mathbf{B}$ is the scalar $\mathbf{A} : \mathbf{B} = A_{ik} B_{ik} = \mathrm{Tr}\left(\mathbf{A}.\mathbf{B}^{\mathrm{T}}\right)$, with the dot denoting the inner product and $\mathrm{Tr}\left(.\right)$ the trace operator. The material gradient, viz the gradient with respect to the material coordinate $\mathbf{X}$, is noted $\nabla_{\mathbf{X}}$; the partial derivative of a function $f\left(x\right)$ with respect to its argument is sometimes denoted with the short cut $f_{,x} = \dfrac{\partial f}{\partial x}$. The general notation $\delta_a\left(.\right)$ stands for the variation of a quantity $\left(.\right)$ considering that the quantity (written here as an index) a is fixed in the variation. For a volume Ω, the differential surface element is noted $\partial\Omega$. For any quantity a, the notation $\dot{a}$ denotes the material derivative, viz $\dot{a} := \dfrac{da}{dt}$. The short cuts l.h.s. and r.h.s. respectively stand for left-hand side and right-hand side.

2 Eshelby Stress in a Thermodynamic Framework

It is one of the essential purposes of this work to throw some light on the derivation of Eshelby tensors in a thermodynamic framework, in articulation with the stationnarity conditions of thermodynamic action functionals. We thereby follow the variational route of configurational mechanics mentioned in the introduction.

The choice of a proper thermodynamic potential is tied to the choice of its control variables (Callen, 1985); this potential achieves an extremum (usually a minimum) when these control variables are held fixed over a suitable RVE (representative volume element); accordingly, it plays the role of a (generalized) potential energy.

The picture differs when the RVE is heterogeneous, since the surface conditions intervene as boundary conditions, and further determine the proper choice of the control variables. For a RVE, there are a priori no well defined boundary conditions (unless it is considered as homogeneous), and one is then free to chose any set of control variables, in coherence with the physical nature of the phenomena occurring within the RVE, and according to the exchanges of the RVE with its surrounding. Those control variables can for instance be selected considering an external reservoir characterized by fixed values of some parameters (extensive or intensive). For instance, considering a RVE immersed in a reservoir imposing both a constant volume and entropy, the internal energy of this RVE is minimum at equilibrium. The various potentials are usually mutually related via Legendre transformations, which prove the adequate mathematical tool to interrelate potentials together with their control variables in a coherent manner.

2.1 Reminder of Thermodynamics: Gibbs and Gibbs-Duhem Relations

In order to set the stage, and following the axioms of classical thermodynamics as stated in (Callen, 1985), see also Muschik (1993), Wilmanski (1996), let assume the existence of a functional $E = E(V\mathbf{F}, S, \mathbf{N})$, called the internal energy, being extensive with respect to its arguments, those arguments reflecting the different forms of energy:

- Mechanical energy (choosing the transformation gradient $\mathbf{F}$ as a kinematic variable, with a nearby constant volume V);

- Calorific (or heat) energy, represented by the total entropy S ;

- Chemical energy, represented by the number of moles $\mathbf{N} = \{N_k, k = 1..n\}$ of the various species (here synthesized in vector format).

The extensity of E (homogeneity of degree one) expresses as (Euler's theorem):

$$E(\lambda V\mathbf{F}, \lambda S, \lambda \mathbf{N}) = \lambda E(V\mathbf{F}, S, \mathbf{N}), \quad \forall \lambda \in \mathbb{R} \qquad (2.1)$$

Deriving (2.1) with respect to λ at $\lambda = 1$ leads to the Euler's identity

$$\frac{\partial E}{\partial(VF)} : (VF) + \frac{\partial E}{\partial S} S + \frac{\partial E}{\partial N} N = E(VF, S, N) \Rightarrow$$

$$E(VF, S, N) = T(VF, S, N) : (VF) + \theta(VF, S, N) S + \mu_k(VF, S, N).N_k \qquad (2.2)$$

wherein the intensive quantities conjugated to the independent extensities have been introduced: the first Piola-Kirchhoff stress tensor $T(VF, S, N)$, the temperature $\theta(VF, S, N)$, and the chemical potentials $\mu_k(VF, S, N)$ of the species labeled by the index k, each of them being a function of the control variables (F, S, N). It is important to note that the intensive nature of these variables (in the sense they are independent of the volume) means they satisfy the following relationships

$$T(\lambda VF, \lambda S, \lambda N) = \lambda^0 T(VF, S, N) \equiv T(VF, S, N)$$

and similarly for both functions $\theta(VF, S, N)$ and $\mu_k(VF, S, N)$.

Accounting for the definition of intensities associated to the arguments of E then leads to the fundamental Gibbs relation

$$dE = T : d(VF) + \theta dS + \mu_k dN_k$$

Gibbs relation can be alternatively written for the energy density $e = E/V$ as

$$de = T : dF + \theta ds + \mu_k dn_k \qquad (2.3)$$

introducing therein s, n_k the entropy density and mole concentration respectively.

The Gibbs relations lead to the state laws

$$\frac{\partial e}{\partial F} = T; \quad \frac{\partial e}{\partial s} = \theta; \quad \frac{\partial e}{\partial n_k} = \mu_k \qquad (2.4)$$

The differentiation of Euler's identity (2.2) – dividing by the volume V - further leads to the well-known Gibbs-Duhem relation (Callen, 1985)

$$F : dT + s d\theta + n_k.d\mu_k = 0 \qquad (2.5)$$

traducing the mutual adjustment of the intensive variables along the thermodynamic equilibrium path followed by the system. Observe that due to Gibbs-Duhem relation, the internal energy density can be integrated from (2.3) (up to a constant term) as

$$e = T : F + \theta s + \mu_k n_k$$

Previous decomposition underlines the fact that the mechanical contribution of the internal energy incorporates both the strain energy density $W_0 = W_0(\mathbf{F}) := \int \mathbf{T} : d\mathbf{F}$ and the complementary strain energy density $W_0^c(\mathbf{T}) := \int \mathbf{F} : d\mathbf{T}$ (with the representation $\mathbf{F} = \mathbf{F}(\mathbf{T})$ implicitly assumed in the last integral).

Both the Gibbs and Gibbs-Duhem relations are at the roots of thermodynamics; Gibbs-Duhem relation expresses the adjustment of the intensive variables during the variation of the extensities.

2.2 Eshelby Stress in the Light of the Thermodynamics of Open Systems

We consider a representative volume element (volume element isolated within a continuum body, much larger compared to the typical size of the heterogeneities, but also much smaller compared to the macroscopic size of the body) as an open heterogeneous system exchanging simultaneously work, mass and heat with its surrounding.

The consideration of the boundary conditions applied to the RVE sets up a link between the nature of the control variables in the context of thermodynamics (those control variables are the arguments of a corresponding thermodynamic potential) and the nature of the fields that obey the imposed boundary conditions: those surface fields acquire the status of control variables per se. Those boundary conditions have to be further accounted for by a proper variational formulation.

In order to set the stage, let define the grand potential in terms of its spatial density

$$\omega = \omega\left(\theta(\mathbf{x},t), \mathbf{F}(\mathbf{x},t), \mu_k(\mathbf{x},t)\right) := \int \mathbf{T} : d\mathbf{F} + \int \mathbf{F} : d\mathbf{T} \equiv W_0(\mathbf{F}) + W_0^c(\mathbf{T}) \tag{2.6}$$

(the listed arguments of ω will be justified later on). The two previous expressions are taken as definitions of the strain energy and complementary strain energy density (quantities $W_0(\mathbf{F})$ and $W_0^c(\mathbf{T})$) respectively; they represent the area below and above the curve $\mathbf{T} = \mathbf{T}(\mathbf{F})$ respectively. Hence, ω expresses as:

$$\omega[\theta, \mathbf{F}, \mu] = \mathbf{T} : \mathbf{F} \tag{2.7}$$

assuming the existence of a strain and stress free reference state. We thereby consider the framework of continuous open systems exchanging not only work (as conceived in the original work of Eshelby), but also mass (transport of chemical species) and heat with their environment.

The grand potential is a quantity used in statistical mechanics, especially for irreversible processes in open systems (see Goodstein, 1975, p. 19), who refers to it as the Landau potential, or the contribution of Chang (2002). Considering the case of homogeneous gases, the grand potential has been set up as the Legendre transform of the internal energy E with respect to the temperature T and the chemical potential μ, viz (keeping the original notations)

$$\Phi_G := E - TS - \mu N \tag{2.8}$$

The more specific choice of internal variables identified to mole numbers shall be adopted here and in the sequel, without however restricting the generality of the presentation.

From the Gibbs-Duhem relation (2.5), the differential of the grand potential is obtained form a straightforward calculation

$$d\omega = -s d\theta + \mathbf{T} : d\mathbf{F} - n_k d\mu_k \tag{2.9}$$

resulting in the thermodynamic relations

$$\frac{\partial \omega}{\partial \theta} = -s; \quad \frac{\partial \omega}{\partial \mathbf{F}} = \mathbf{T}; \quad \frac{\partial \omega}{\partial \mu_k} = -n_k \tag{2.10}$$

highlighting $\theta, \mathbf{F}, \mu_k$ as the arguments of ω. Hence, relations (2.10) show that the grand potential is the extension to the tensorial case of the original definition (2.8) in the case of homogeneous gases. It further represents the partial Legendre transform of the internal energy with respect to the temperature and the number of moles (here the internal variable α, see (Callen, 1985, pp. 146-148). The total Legendre transform of the internal energy would give the zero potential, see (Callen, 1985, p. 148, equ. (5.60)).

Remark: the contributions on the r.h.s. of $d\omega$ in (2.9) are successively the quasi-static heat flux, the incremental work of the internal stresses and the so-called quasi-static chemical work (Callen, 1985, p. 36).

The grand potential is next involved to build a weak form of thermodynamic equilibrium. From the static equilibrium equation

$$\nabla_X . \mathbf{T} + \mathbf{f}_0 = \mathbf{0}$$

with $\mathbf{f}_0$ the body forces, the definition of the transformation gradient

$$\mathbf{F} := \nabla_X \mathbf{x}$$

and an integration by part, the integral of ω on a control volume V is related to contributions involving the applied loading (body forces and surface tractions $\mathbf{T.N}$):

$$\int_V \omega dV = \int_V \mathbf{f}_0 . \psi dV + \int_{\partial V} (\mathbf{T.N}) . \psi dS \tag{2.11}$$

with ψ a virtual kinematically admissible position field. The boundary ∂V can be partitioned such as to evidence a control in the placement on a portion $S_{0\psi}$

and a control in stress (or rather in traction) on a complementary portion S_{0t}, (with $S_{0\psi} \cap S_{0t} = \varnothing$), according to

$$\int_V \omega dV = \int_V \mathbf{T}:\mathbf{F}dV = \left\{ \int_V \mathbf{f}_0.\psi dV + \int_{S_{0t}} \left(\mathbf{T}^d.\mathbf{N}\right).\psi dS \right\} + \int_{S_{0\psi}} \left(\mathbf{T}.\mathbf{N}\right).\psi^d dS \equiv W_T + W_\psi$$

(2.12)

involving the controlled (prescribed) stress and displacements $\mathbf{T}^d, \psi^d$ respectively, introducing therein the works of the imposed body forces and tractions W_T and the work of the imposed placements, viz the quantity $W_\psi = \int_{S_{0\psi}} \left(\mathbf{T}.\mathbf{N}\right).\psi^d dS$.

The Gibbs-Duhem relation allows expressing one of the intensive variables in the triplet $\left(\theta, \mathbf{T}, \mu_k\right)$ vs. the two remaining variables. Using (2.11), a Legendre transform of ω versus the work terms gives a new potential (this constitutes a Legendre transformation, whereby the fixed surface fields have been eliminated from the set of arguments of the resulting potential), which identically vanishes, called for this reason the zero potential, defined as:

$$\Omega\left[\theta, \psi, \mu_k\right] := \int_{V_0} \omega dV - \left\{ \int_{V_0} \mathbf{f}_0.\psi dV + \int_{S_{0t}} \left(\mathbf{T}^d.\mathbf{N}\right).\psi dS \right\} - \int_{S_{0\psi}} \left(\mathbf{T}.\mathbf{N}\right).\psi^d dS \equiv 0$$

(2.13)

Observe that the zero potential constructed thereabove is the tensorial generalization of the zero potential, accounting for the boundary conditions over the (heterogeneous) RVE.

The zero potential is constantly nil, hence it is extremal, viz

$$\delta\Omega\left[\theta, \psi, \mu_k\right] = 0 \tag{2.14}$$

for the solution field at the thermodynamic equilibrium of the system, and whatever the material constitutive law be. Thereby, a generalized potential energy in a thermodynamic framework has been set up, allowing for the consideration of the thermal and chemical forms of energy, in addition to the mechanical energy.

Observe that the stress has been substituted by the temperature and chemical potential as new control variables: it is thus not possible to control the stress, and essential boundary conditions on the portion of surface S_t have to be prescribed in the formulation of the associated extremum principle, namely

$$\mathbf{T}.\mathbf{N} = \mathbf{t}^d \text{ on } S_{0t} \tag{2.15}$$

The variation of the three fields functional $\Omega[\theta, \psi, \mu_k]$ in (2.13) is next evaluated (the integration volume and surface are fixed), with the body forces $\mathbf{f}_0$ and surface tractions $\mathbf{t}^d$ acting in the fixed volume and on the fixed part of the boundary $S_{0\psi}$ and S_{0t} respectively. The displacements are held fixed on the part of the boundary $S_{0\psi}$, equal to ψ^d. Note that this constitutes a mixed (three fields) principle in a thermodynamic sense (due to the Gibbs-Duhem relation), albeit it becomes a two-field principle when the set of variables (θ, μ_k) is substituted by the stress variable $\mathbf{T}$.

The material variation (at fixed material position: the index $\mathbf{X}$ is dropped in the variation, hence $\delta \equiv \delta_X$) of $\Omega[\theta, \psi, \mu_k]$ is obtained as

$$\delta\Omega[\theta, \psi, \mu_k] = \int_{V_0} \mathrm{Div}\left(-\Sigma - \left(s\theta + n_k\mu_k\right)\mathbf{I}\right).\delta\mathbf{X}dX - \int_{V_0}\left(\theta\mathrm{Div}\left(s\mathbf{I}\right) + \mu_k\mathrm{Div}\left(n_k\mathbf{I}\right)\right).\delta\mathbf{X}dX +$$

$$+\int_{V_0}\left(\partial_X W_0\right)_{\mathrm{expl}}.\delta\mathbf{X}dX + \int_{V_0} \mathbf{F}^T.\mathbf{f}_0.\delta\mathbf{X}dX + \int_{S_{0t}} \frac{\partial\omega}{\partial\nabla_X\psi}.\mathbf{N}.\delta\psi dS - \int_{S_{0t}} \mathbf{t}^d.\delta\psi dS - \int_{S_{0\psi}} \delta(\mathbf{T}.\mathbf{N}).\psi^d dS$$

$$(2.16)$$

introducing therein the mechanical Eshelby stress (Eshelby, 1957; Markenscoff and Gupta, 2006)

$$\Sigma = \omega\mathbf{I} - \mathbf{F}^T.\mathbf{T} \tag{2.17}$$

Observe that this Eshelby stress incorporates the density of the grand potential, sum of the strain energy density $W_0(\mathbf{F})$ and of the complementary energy density $W_c(\mathbf{T})$; hence it differs from the classical version, which relies on $W_0(\mathbf{F})$ exclusively. In deriving (2.16), (2.17), one has used the following expression of the material variation

$$\delta\omega = \delta\omega(\theta, \mathbf{F}, \mu_k) \equiv \frac{\partial\omega}{\partial\nabla_X\psi} : \delta(\nabla_X\psi) + \frac{\partial\omega}{\partial\theta}\delta\theta + \frac{\partial\omega}{\partial\mu_k}\delta\mu_k$$

The first contribution on the r.h.s. has been integrated by part, and the variations therein have been expressed successively as:

$$\delta_x\psi + \mathbf{F}.\delta_x\mathbf{X} = \mathbf{0}; \quad \delta\theta = \nabla_X\theta.\delta\mathbf{X}; \quad \delta\mu_k = \nabla_X\mu_k.\delta\mathbf{X}$$

Adopting the synthetic notation $\delta\Omega_{\mathrm{vol}}$ for the volumetric contributions in the variation $\delta\Omega_{\mathrm{vol}}$, the stationnarity of $\Omega[\theta, \psi, \mu_k]$ then leads to the balance law (a necessary condition)

$$\delta\Omega_{\mathrm{vol}} = 0 \Rightarrow \mathrm{Div}\Xi + \left(\theta\nabla_X s + \mu_k\nabla_X n_k\right) + \mathbf{F}^T.\mathbf{f}_0 = \mathbf{0} \tag{2.18}$$

introducing therein the following energy momentum tensor in a thermodynamic context

$$\Xi := -\Sigma - \left(s\theta + n_k \mu_k\right)\mathbf{I} \tag{2.19}$$

This generalized Eshelby stress includes a first contribution recognized as the mechanical Eshelby stress (accounting for the mechanical energy contribution of the internal energy, as reflected by (2.17)) and a second contribution accounting for the thermal and chemical forms of energy.

The Eshelby tensor in (2.19) is similar in form to the Eshelby material tensor elaborated in Maugin (2006a), considering the scheme of the classical entropy-heat flux relation, with the contributions of the different form of energies explicited. Eshelby material stresses have been further derived by the same author considering dissipative solid materials described by a diffusive internal variable, Maugin (2006b), letting the Helmholtz energy density additionally depend upon the gradient of the internal variable. The reader is also referred to the works of Epstein and Maugin (2000), Maugin (1993) and Eshelby (1970) related to those tensors.

The balance law (2.18) is supplemented by the natural boundary conditions

$$t^d = \frac{\partial \omega}{\partial \nabla_X \psi} . \mathbf{N} \tag{2.20}$$

3 Eshelby Tensor and Gibbs-Duhem Constraint as a Transversality Condition

The goal of this section is to recast the classical Gibbs-Duhem condition (see Callen (1985)) into a more general tensorial form, derived as a transversality condition of a thermodynamic action integral. Consider the Hamiltonian action

$$A := \int_\Omega l(\mathbf{a}, \mathbf{u}) \, d\Omega \tag{3.1}$$

with the spatio-temporal integration domain $\Omega = V \times \left[t_1, t_2\right]$, involving the Lagrangian density $l(\mathbf{a}, \mathbf{u})$ (per unit volume and per unit time), noting here $(\mathbf{a}, \mathbf{u}) \equiv \left(a^i, u^j\right)$ the set of independent and dependent variables respectively (in the sense that each component u^j is a function of the coordinates a^i), written as contravariant vectors, arguments of the density $l(\mathbf{a}, \mathbf{u})$. The general notation V for the volume is adopted in the sequel.

Observe that due to its status as a thermodynamic potential, the density $l(\mathbf{a}, \mathbf{u})$ cannot depend explicitly upon time and space (here the chosen parameterization $\mathbf{a}$), hence $l(\mathbf{a}, \mathbf{u}) \equiv l(\mathbf{u})$, which means that the lagrangian density depends on the sole fields.

The Lagrangian density may be selected as the rate of the density of any thermodynamic potential, in coherence with a proper choice of thermodynamic control variables: selecting for instance the rate of the internal energy density in (2.4), the stationnarity of the Hamilton action in (4.1) delivers as necessary condition

$$A := \int_{\Omega} \dot{e}\left(\mathbf{F}, s, n_k\right) d\Omega \rightarrow \delta A = 0 \Rightarrow \forall x \in \left\{\mathbf{F}, s, n_k\right\}, \quad \frac{d}{dt}\left(\frac{\partial \dot{e}}{\partial \dot{x}}\right) - \frac{\partial e}{\partial x} = 0$$

with x denoting any of the three variables $\mathbf{F}, s, n_k$. Those three equations express the state laws (2.4) in rate form, involving the Hessian of the internal energy (Magnenet at al., 2007):

$$\begin{pmatrix} \dot{\mathbf{T}} \\ \dot{\theta} \\ \dot{\mu}_k \end{pmatrix} = \begin{pmatrix} e_{,FF} & e_{,Fs} & e_{,F\mu_k} \\ e_{,sF} & e_{,ss} & e_{,s\mu_k} \\ e_{,n_k F} & e_{,n_k\theta} & e_{,n_k n_k} \end{pmatrix} \cdot \begin{pmatrix} \dot{\mathbf{F}} \\ \dot{s} \\ \dot{n}_k \end{pmatrix} \tag{3.2}$$

In the same vein, and drawing the parallel with (3.2), using the rate of the grand potential as a lagrangian viz

$$\dot{\omega} = -s\dot{\theta} + \mathbf{T} : \dot{\mathbf{F}} - n_k \dot{\mu}_k$$

the following Euler equations are obtained

$$-\dot{s} = \frac{\partial \dot{\omega}}{\partial \theta} \equiv -\frac{\partial s}{\partial \theta}\dot{\theta} + \frac{\partial \mathbf{T}}{\partial \theta} : \dot{\mathbf{F}} - \frac{\partial n_k}{\partial \theta}\dot{\mu}_k \equiv \omega_{,\theta\theta}\dot{\theta} + \omega_{,\theta F} : \dot{\mathbf{F}} + \omega_{,\theta\mu_k}\dot{\mu}_k$$

$$\dot{\mathbf{T}} = \frac{\partial \dot{\omega}}{\partial F} \equiv -\frac{\partial s}{\partial F}\dot{\theta} + \frac{\partial \mathbf{T}}{\partial F} : \dot{\mathbf{F}} - \frac{\partial n_k}{\partial F}\dot{\mu}_k \equiv \omega_{,F\theta}\dot{\theta} + \omega_{,FF} : \dot{\mathbf{F}} + \omega_{,F\mu_k}\dot{\mu}_k \tag{3.3}$$

$$-\dot{n}_k = \frac{\partial \dot{\omega}}{\partial \mu_k} \equiv -\frac{\partial s}{\partial \mu_k}\dot{\theta} + \frac{\partial \mathbf{T}}{\partial \mu_k} : \dot{\mathbf{F}} - \frac{\partial n_k}{\partial \mu_k}\dot{\mu}_k \equiv \omega_{,\mu_k\theta}\dot{\theta} + \omega_{,\mu_k F} : \dot{\mathbf{F}} + \omega_{,\mu_k\mu_k}\dot{\mu}_k$$

Those necessary conditions are fully equivalent to the state equations in rate form involving the Hessian of the grand potential, which also result by taking the rate of relations (2.10).

The variation of the functional A under the action of a Lie group (continuous group of transformations) with parameter μ is next obtained (Olver, 1993) as the sum of a volumetric contribution $\delta A_{vol} = \int_V \mathbf{Q}.EL\left[1\left(\mathbf{u}\right)\right] dV$, involving the Euler operator $EL\left(.\right)$ and the characteristic of the generator vector field $\mathbf{Q}$, and the surface contribution

$$\delta A_{surf} = \mu \int_{\partial V}\left(1\xi^i + \left(\phi^k - u_{,j}^k \xi^j\right)\frac{\partial 1}{\partial u_{,j}^k}\right) N_i d\left(\partial V\right) \tag{3.4}$$

keeping the generic notations $u^k_{,j}$, which encompass both the spatial and temporal derivatives of the dependent variables u^k. The components ξ (resp. ϕ) denote the horizontal (resp. vertical) variations of the independent (resp. dependent) variables, such that $\mu\xi = \delta\mathbf{X}$ and $\mu\phi = \delta\mathbf{u}$.

On the optimal trajectory, both the volumetric and the surface contributions to δA have to vanish, viz

$$\delta A = 0 \rightarrow \left| \begin{aligned} \delta A_{vol} &= 0 \\ \delta A_{surf} &= 0 \end{aligned} \right. \tag{3.5}$$

Adopting further the elegant and compact viewpoint of differential geometry (Cartan, 1977; Ganghoffer, 2007), the previous variation of the action integral in (3.1) can be equivalently written in the more compact differential form

$$L_X \varpi = i_X d\varpi + d\left(i_X \varpi\right) \tag{3.6}$$

that allows a condensed writing of the

Noether's theorem (differential version): under the conditions $L_X \varpi = 0$ (invariance of $\int_\Omega \varpi \equiv \int_\Omega ldX$ by the group generated by X) and $i_X d\varpi = 0$ (the Euler equations are identically satisfied), the following conservation law is obtained:

$$d\left(i_X \varpi\right) = 0 \tag{3.7}$$

Thereby, in the language of differential geometry, and as a consequence of the optimality conditions of Jacobi action, the quantity $i_X \varpi$, representing the interior product of the one form $\varpi = ldX$ by the vector field $\mathbf{X}$, is conserved along the optimal paths followed by the media described by the Lagrangian density L. This condition represents a strong form of the stationnarity condition of the action integral S, in the sense of the implication

$$i_X d\varpi = 0 \Rightarrow \delta \int_C \varpi = 0$$

From a mechanical point of view, this incremental formulation of a least action principle is in agreement with the differential writing of the hyperelastic constitutive law ($d\mathbf{T} = \dfrac{\partial W_0}{\partial \mathbf{F}} : d\mathbf{F}$); it is accordingly best adapted to the formulation of a least action principle.

As a necessary optimality condition of the Hamiltonian action (3.1), the vanishing of the surface contribution (3.4), viz the condition $\delta A_{surf} = 0$, renders for

fixed values of the fields at the boundary (in the case of a purely horizontal variation: $\phi^k = 0 \leftrightarrow \delta u^k = 0$) the condition:

$$d\left(i_X \varpi\right) = 0 \Rightarrow \left(l\delta_i^j - u_{,j}^k \frac{\partial l}{\partial u_{,i}^k}\right) N_j = 0 \tag{3.8}$$

This condition is further explicited, adopting the lagrangian density
$l \equiv \dot{\omega} = \dfrac{d\omega}{dt}$, with $\omega = \omega\left(\theta, \mathbf{F}, \mu_k\right)$: one works out the term

$$l\delta_i^j - u_{,j}^k \frac{\partial l}{\partial u_{,j}^k} \equiv \dot{\omega}\delta_i^j - \left(-\dot{\theta}s\delta_i^j - n_k\dot{\mu}_k\delta_i^j + T_{ik} : \dot{F}_{jk}\right) \tag{3.9}$$

$$\leftrightarrow l\mathbf{I} - \dot{\mathbf{u}} : \frac{\partial l}{\partial \dot{\mathbf{u}}} \equiv \dot{\omega}\mathbf{I} - \left(-\dot{\theta}s\mathbf{I} - n_k\dot{\mu}_k\mathbf{I} + \mathbf{T}.\dot{\mathbf{F}}^T\right)$$

From (2.7), we further express the Lagrangian density

$$l = \dot{\omega} = \dot{\mathbf{T}} : \mathbf{F} + \mathbf{T} : \dot{\mathbf{F}} \tag{3.10}$$

which inserted into (3.8) gives (in tensorial format)

$$l\mathbf{I} - \dot{\mathbf{u}} : \frac{\partial l}{\partial \dot{\mathbf{u}}} = \left(\mathbf{T} : \dot{\mathbf{F}}\right)\mathbf{I} - \mathbf{T}.\dot{\mathbf{F}}^T + \left(\dot{\mathbf{T}} : \mathbf{F}\right)\mathbf{I} + \left(\dot{\theta}s + n_k\dot{\mu}_k\right)\mathbf{I} \tag{3.11}$$

Hence, the necessary condition (3.7) becomes

$$\left(\mathbf{T} : \dot{\mathbf{F}}\right)\mathbf{N} - \left(\mathbf{T}.\dot{\mathbf{F}}^t\right).\mathbf{N} + \left(\dot{\mathbf{T}} : \mathbf{F}\right)\mathbf{N} + \left(\dot{\theta}s + n_k\dot{\mu}_k\right)\mathbf{N} = \mathbf{0} \text{ on } \partial\Omega \tag{3.12}$$

which can be alternatively written (see (3.9))

$$\left(\dot{\omega}\mathbf{I} - \mathbf{T}.\dot{\mathbf{F}}^t\right).\mathbf{N} + \left(\dot{\theta}s + n_k\dot{\mu}_k\right)\mathbf{N} = \mathbf{0} \tag{3.13}$$

Thereby, a generalized tensorial-like Gibbs-Duhem condition resulting from
the transversality condition of the Jacobi action (3.1) has been obtained.

4 Summary and Perspectives

As a main thrust of this contribution, the notion of energy momentum tensor has
been related to the thermodynamics of open continuum bodies exchanging work,
heat and chemical species (hence mass) with their surrounding, in articulation
with the calculus of variations, and especially the so-called transversality condition (considering a domain with movable boundaries). Thereby, the mathematical
structure of thermodynamics has been exploited, namely Gibbs and Gibbs-Duhem
conditions have been used as its two basic pillars. The Lagrangian density
entering the Hamiltonian action has been identified to the material derivative of

a thermodynamic potential, rendering the state laws in rate form as the Euler equations.

An Eshelby stress including mechanical, thermal and chemical contributions has been constructed, starting from a weak form of the equilibrium equations involving the grand potential. Further thermodynamic-like energy-momentum tensors may be constructed in a similar manner, starting from a specific choice of a thermodynamic potential (or rather its rate). All these potentials are equivalent to each other, in the sense that they are mutually related by a Legendre-Fenchel transform; amongst all those potentials, the grand potential plays however a privileged role, in the sense it allows formulating a universal extremum principle, valid for a general hyperelastic constitutive law.

The Gibbs-Duhem condition in tensorial format has further been obtained from the optimality conditions, as the transversality condition associated to the Hamiltonian action.

As a final remark, observe that the Eshelby stress appears in the domain variation of the Hamiltonian action (the surface contribution (3.4)); thereby, it enters into the expression of the material forces (of a thermodynamic nature) that trigger the domain variation in the context of open continuum solid bodies exchanging both mass, heat, and work with their surrounding.

References

Eshelby, J.D.: The force on an elastic singularity. Philosophical Transactions of the Royal Society of London. Series A, Mathematical and Physical Sciences 244(877), 87–112 (1951)

Maugin, G.A.: Material Inhomogeneities in Elasticity. Chapman et Hall, Boca Raton (1993)

Maugin, G.A.: Material forces: concepts and applications. ASME Appl. Mech. Rev. 48, 213–245 (1995)

Gurtin, M.E.: Configurational forces as basic concepts of Continuum Physics. Springer, Berlin (2000)

Gross, D., Kolling, S., Mueller, R., Schmidt, I.: Configurational forces and their application in solid mechanics. Eur. J. Mech. A/ Solids 22, 669–692 (2003)

Kuhl, E., Steinmann, P.: Material forces in open system mechanics. Comput. Methods Appl. Mech. Engrg. 193, 2357–2381 (2004)

Lubarda, V.A., Markenscoff, X.: Complementary energy release rates and dual conservation integrals in micropolar elasticity. J. Mech. Phys. Solids 55(10), 2055–2072 (2007)

Menzel, A., Steinmann, P.: On configurational forces in multiplicative elastoplasticity. Int. J. Solids Struct. 44(13), 4442–4471 (2007)

Fagerström, M., Larsson, R.: Approaches to dynamic fracture modelling at finite deformations. J. Mech. Phys. Solids 56(2), 613–639 (2008)

Agiasofitou, E.K., Kalpakides, V.K.: The concept of a balance law for a cracked elastic body and the configurational force and moment at the crack tip. Int. J. Engng. Sci. 44(1-2), 127–139 (2006)

Steinmann, P.: On boundary potential energies in deformational and configurational mechanics. J. Mech. Phys. Solids 56(3), 772–800 (2008)

Edelen, D.G.B.: Aspects of variational arguments in the theory of elasticity: fact and folklore. Int. J. Solids Struct. 17, 729–740 (1981)

Honein, T., Chien, N., Herrmann, G.: On conservation laws for dissipative systems. Phys. Lett. A. 155, 223–224 (1991)

Chien, N., Herrmann, G.: Conservation laws for thermo or poroelasticity. J. Applied Mech. 63, 331–336 (1996)

Kalpakides, V.K., Maugin, G.A.: Canonical formulation and conservation laws of thermoelasticity without dissipation. Reports on Math. Phys. 3(54), 371–391 (2004)

Callen, H.B.: Thermodynamics and an introduction to thermostatics. J. Wiley, New York (1985)

Muschik, W.: Fundamentals of non equilibrium thermodynamics. In: Muschik, W. (ed.) Proc. CISM Course and Lectures, Non-equilibrium thermodynamics with applications to solids, vol. 336 (1993)

Wilmanski, K.: Thermomechanics of Continua. Springer, Heidelberg (1996)

Magnenet, V., Rahouadj, R., Ganghoffer, J.F., Cunat, C.: Continuous symmetries and constitutive laws of thermo-elasto(viscoplastic materials within a thermodynamical framework of relaxation. Part I: formal aspects. Int. J. Plasticity 23(1), 87–113 (2007)

Goodstein, D.: States of Matter. Dover Phoenix Editions (1975)

Chang, L.J.: Thermal Physics - Entropy and Free Energies, ch. 5. World Scientific, New Jersey (2002)

Eshelby, J.D.: The Determination of the Elastic Field of an Ellipsoidal Inclusion. In: Proceedings of the Royal Society of London, vol. A241, p. 376 (1957)

Markenscoff, X., Gupta, A.: Collected Works of J. D. Eshelby, Mechanics of Defects and Inhomogeneities. Springer, Heidelberg (2006)

Maugin, G.A.: On canonical equations of continuum thermomechanics. Mech. Res. Comm. 33, 705–710 (2006a)

Maugin, G.A.: On the thermomechanics of continuous media with diffusion and /or weak nonlocality. Arch. Appl. Mech. 75, 723–738 (2006b)

Epstein, M., Maugin, G.A.: Thermomechanics of volumetric growth in uniform bodies. Int. J. Plasticity 16(7-8), 1, 951–978 (2000)

Eshelby, J.D.: Energy relations and the energy—momentum tensor in continuum mechanics. In: Kanninen, M.F., Adler, W.F., Rosenfield, A.R., Jaffee, R.I. (eds.) Inelastic Behaviour of Solids, pp. 77–115. McGraw-Hill, New York (1970)

Basar, Y., Weichert, D.: Nonlinear continuum mechanics of solids. Springer, Heidelberg (2000)

Cartan, H.: Cours de calcul différentiel. Hermann (ed.) (1977)

Ganghoffer, J.F.: Differential geometry, least action principles and irreversible processes. Rend. Sem. Mat. Univ. Pol. Torino 65(2), 43–73 (2007)

Olver, P.J.: Applications of Lie groups to Differential Equations. In: Graduate Texts in Mathematics, vol. 107. Springer, New York (1993)

Nonlinear Hyperbolic Equations and Linear Heat Conduction with Memory

Sandra Carillo

Abstract. The model of a rigid heat conductor with memory is considered. Specifically, in the one-dimensional case, a connection, via Cole-Hopf Transformation, between the linear integro-differential evolution equation which describes heat conduction with memory and a nonlinear partial integro-differential equation of hyperbolic type is established. Notably, when the heat conductor is homogeneous, as well as when the homogeneity hypothesis is removed, the differential operator of the transformed nonlinear partial differential equation is of hyperbolic type.

1 Introduction

Heat conduction problems are here considered within the framework of *materials with memory*. Indeed, the introduction of memory effects to overcome the infinite speed of propagation of heat, exhibited by the classical linear heat equation, goes back to Cattaneo [2] who suggested a new generalized Fourier's law which linearly relates the heat flux, its time derivative and the temperature-gradient. Since then, the model of heat conduction with memory has been widely studied to investigate the behaviour of those materials in the case when their thermodynamical status depends on time not only through the present time, but also through its past history and, hence, memory effects cannot be neglected.

The model here referred to takes its origin in the work by Coleman [5] and, later on, by Gurtin and Pipkin [10]. Heat conduction in a fading memory material is studied by Giorgi and Gentili [8]. In particular, the integro-differential equation to start from is that one obtained by Fabrizio, Gentili and Reynolds [7], who were concerned about the thermodynamics of a rigid homogeneous heat conductor with memory. Furthermore, hyperbolicity of Volterra integro-differential model of heat

Sandra Carillo
Dipartimento di Metodi e Modelli Matematici per le Scienze Applicate,
University of Rome "La Sapienza", Via A. Scarpa 16, I-00161 Rome, Italy
e-mail: `carillo@dmmm.uniroma1.it`, `sandra.carillo@uniroma1.it`

J.-F. Ganghoffer, F. Pastrone (Eds.): Mech. of Microstru. Solids 2, LNACM 50, pp. 63–70.
springerlink.com   © Springer-Verlag Berlin Heidelberg 2010

conduction with memory has been observed by Gurtin and Pipkin [10] and, later by Davis [6] and Belleni-Morante [1].

The study of the linear integro-differential evolution equation is here attacked under a perspective which is different from previous works in the wide literature. Indeed, the aim is to show that the model equation under investigation is related to an hyperbolic nonlinear wave type equation, via a Cole-Hopf Transformation, a special case of the much wider class of Bäcklund Transformations. The latter represents a powerful tool in investigating nonlinear differential equations both in revealing structural properties as well as in finding new solutions to assigned problems. Notably, when two initial boundary value problems are connected via a Bäcklund Transformation, the same link between their solutions is established. Hence, solutions to a problem can be constructed on transforming the corresponding ones of the other problem, linked via Bäcklund Transformation. This, is exactly the framework in which, independently, Cole [4] and Hopf [11] solved an initial value Burgers problem recognizing that Burgers equation is related via a Bäcklund Transformation, thereafter known as Cole-Hopf transformation, to the classical linear heat equation. Accordingly, in [9] the solution of an infiltration in soils problem, modeled via a Burgers initial boundary value problem, is explicitly obtained in terms of complementary error functions.

An extensive bibliography concerning applications of Bäcklund Transformations in solving boundary and initial value problems is comprised in Rogers and Shadwick [15], and, more recently, in Rogers and Ames [16].

The material is organized as follows. Section 2 is devoted to a brief introduction of the model under investigation to show how the integro-differential evolution problem of interest is obtained. Section 3 concerns the connection, via Cole-Hopf Trasformation, of the linear Gurtin-Pipkin integro-differential evolution problem with a nonlinear *wave type* one. The closing Section 4 comprises some remarks and connections with other works in the literature.

2   Evolution Problems in Thermodynamics with Memory

This section is concerned about a brief introduction to the model which gives rise to an integro-differential equations to describe heat conduction in a rigid media when *memory* effects occur. Here, for sake of simplicity, the internal energy, denoted by e, is assumed to be proportional to the relative temperature, thus

$$e(\mathbf{x},t) = \alpha_0 v(\mathbf{x},t) \ , \tag{1}$$

where t denotes the time variable, and $v := \theta - \theta_0$ the temperature difference with respect to a fixed reference temperature θ_0, and α_0 represents the *energy relaxation function*, here assumed to be constant, $\mathbf{x} \in \Omega \subset \mathbb{R}^3$ denotes the position within Ω, the bounded closed set in $\mathbb{R}^3$ which represents the configuration domain of the conductor. The heat flux $\mathbf{q} \in \mathbb{R}^3$ is assumed to satisfy the constitutive equation

$$\mathbf{q}(\mathbf{x},t) = -\int_0^\infty k(\mathbf{x},\tau)\nabla\theta(\mathbf{x},t-\tau)\,d\tau \ , \tag{2}$$

where, since θ_0 denotes a constant, $\nabla v = \nabla\theta$ is the temperature-gradient and $k(\mathbf{x},\tau)$ the heat flux relaxation function. The latter is subject to the analytical restrictions required by Giorgi and Gentili in [8] and Fabrizio, Gentili and Reynolds [7] to guarantee the physical admissibility of the model.

The space-time domain wherein the unknown function v, which represents the temperature, is defined is $Q_T = \Omega \times (0,T) \subset \mathbb{R}^3 \times \mathbb{R}$, where $\Omega \subset \mathbb{R}^3$ denotes the heat conductor configuration domain. The evolution problem reads:

$$v_t = -\nabla \cdot q(\mathbf{x},t) + r(\mathbf{x},t) \tag{3}$$

$$\mathbf{q}(\mathbf{x},t) = +\int_0^\infty \dot{k}(\mathbf{x},s)\overline{\nabla v}^t(\mathbf{x},s)\,ds \tag{4}$$

where $\dot{k}$ denotes the partial derivative of k with respect to the time variable, while $r(\mathbf{x},t)$ denotes the heat supply and

$$\overline{\nabla v}^t(\mathbf{x},s) := \int_{t-s}^t \nabla v(\mathbf{x},\tau)\,d\tau \ , \tag{5}$$

the integrated history of the temperature-gradient; hence, on substitution of the expression of the heat flux q in (3), the evolution equation reads

$$v_t = -\nabla \cdot \int_0^\infty \dot{k}(\mathbf{x},s)\overline{\nabla v}^t(\mathbf{x},s)\,ds + r(\mathbf{x},t) \ . \tag{6}$$

According to [3], the evolution problem can be, equivalently, written under the form

$$v_t(\mathbf{x},t) = \nabla \cdot \int_0^t k(\mathbf{x},t-\tau)\nabla v(\mathbf{x},\tau)d\tau - \nabla \cdot \bar{\mathbf{I}}^0(\mathbf{x},t) \tag{7}$$

where $\bar{\mathbf{I}}^0(\mathbf{x},t)$, defined via

$$\bar{\mathbf{I}}^0(\mathbf{x},t) := \int_0^\infty \dot{k}(\mathbf{x},t+\tau)\overline{\nabla v}^0(\mathbf{x},\tau)d\tau \ , \tag{8}$$

is assigned together with the initial and boundary conditions. When the heat supply $r(\mathbf{x},t)$ is assumed to be in *gradient form*, namely $r(\mathbf{x},t) := -\nabla \cdot R(\mathbf{x},t)$, then the initial status as well as the source term, are comprised in

$$\bar{\mathbf{I}}^0(\mathbf{x},t) := \int_0^\infty k'(\mathbf{x},t+\tau)\overline{\nabla v}^0(\mathbf{x},\tau)d\tau + R(\mathbf{x},t)$$

$$\overline{\nabla v}^0(\mathbf{x},t) := \int_{-t}^0 \nabla v(\mathbf{x},s)\,ds \tag{9}$$

while the initial conditions are given when the initial temperature distribution $v_0(\mathbf{x})$, say $v_0 \in C^1[0,L]$, is assigned and boundary conditions are prescribed further to a sufficiently smooth

$$\bar{\mathbf{I}}^{t=0}(\mathbf{x},t) = \bar{\mathbf{I}}^0(\mathbf{x},t), \quad t \in (0,T). \tag{10}$$

Specifically, when an isolated heat conductor is considered, initial and boundary conditions read, in turn

$$v(\mathbf{x},0) = v_0(\mathbf{x}), \tag{11}$$

$$q(\mathbf{x},t) \cdot n(\mathbf{x}) = 0, \quad \forall t \geq 0 \tag{12}$$

where n denotes the outward unit normal t_0 the smooth boundary $\partial\Omega$. Notably, according to (9), it is required to assign the thermal history only in the finite time interval $(-t,0)$, $t \in [0,T]$.

The existence and uniqueness of the solution to such a problem has been already proved in [3] on application of variational methods. Here, in the following Section, the hyperbolic behaviour of the evolution equation (7), in $1+1$-dimensions, is obtained on recognizing that it is connected, through a Cole-Hopf transformation, to a nonlinear hyperbolic equation.

3   Cole-Hopf Transformation

This Section is concerned about how to relate the equation under investigation to a nonlinear hyperbolic-type equation. For sake of simplicity, the case when the heat conductor is 1-dimensional is considered. Accordingly, the integro-differential evolution equation (7) reads

$$v_t = \left[\int_0^t k(x,t-\tau)v_x(x,\tau)\,d\tau - \bar{\mathbf{I}}^0(x,t) \right]_x, \tag{13}$$

where now $x \in \Omega = [0,L] \subset \mathbb{R}$. In addition, the heat flux relaxation function is assumed to depend only the time variable t, namely, the case of a homogeneous rigid heat conductor is considered. Then, the adopted thermodynamical assumptions [8] [7], imply that

$$k(t) = k_0 + \int_0^t \dot{k}(s)\,ds , \tag{14}$$

where $k_0 \equiv k(0)$ represents the initial (positive) value of the heat flux relaxation function, thus termed *initial heat flux relaxation coefficient*. It is further required that

$$\dot{k} \in L^1(\mathbb{R}^+) \cap L^2(\mathbb{R}^+) \quad \textit{and} \quad k \in L^1(\mathbb{R}^+) \tag{15}$$

which imply $k(\infty) := \lim_{t \to \infty} k(t) = 0$. The latter can be physically interpreted recalling that there is no heat flux when, at infinity, the thermal equilibrium is reached.

Hence the evolution equation (13) reduces to

$$v_t = \int_0^t k(t-\tau)v_{xx}(x,\tau)\,d\tau - \bar{\mathbf{I}}^0_x(x,t) , \quad (x,t) \in [0,L] \times \mathbb{R} \tag{16}$$

when the source term is set equal to zero. Partial derivation with respect to t of such an equation shows its hyperbolic type character. Exactly this character is emphasized on showing that (16) connected via a Cole-Hopf Transformation [4, 11]

$$v_x - uv = 0 \tag{17}$$

to a nonlinear hyperbolic equation[1]. Indeed, (17) implies

$$u = D_x \ln v \ , \quad D_x := \frac{\partial}{\partial x} \tag{18}$$

and, on introduction of its inverse operator[2], denoted as D_x^{-1}, defined via

$$D_x^{-1} f(x, \cdot) := \int_0^x f(\xi, \cdot) d\xi \tag{19}$$

$$v(x, \cdot) = e^{D_x^{-1} u(x, \cdot)} \implies v_x(x, \cdot) = u(x, \cdot) e^{D_x^{-1} u(x, \cdot)} \tag{20}$$

and

$$v_t(x, \cdot) = D_x^{-1} u_t(x, \cdot) e^{D_x^{-1} u(x, \cdot)} \tag{21}$$

that, after multiplication of both sides by $e^{-D_x^{-1} u(x,t)}$, gives

$$D_x^{-1} u_t(x,t) = e^{-D_x^{-1} u(x,t)} \int_0^t k(t - \tau) e^{D_x^{-1} u(x,\tau)} [u^2(x, \tau) + u_x(x, \tau)] d\tau - \bar{\mathbf{I}}_x^0(x,t) \ . \tag{22}$$

which can be written also as

$$D_x^{-1} u_t(x,t) = \int_0^t k(t - \tau) e^{D_x^{-1}[u(x,\tau) - u(x,t)]} [u^2(x, \tau) + u_x(x, \tau)] d\tau - \bar{\mathbf{I}}_x^0(x,t) \ . \tag{23}$$

The latter, on application of the operator D_x on the left of both sides, reads

$$u_t = \int_0^t k(t - \tau) \left\{ e^{D_x^{-1}[u(x,\tau) - u(x,t)]} [u^2(x, \tau) + u_x(x, \tau)] \right\}_x d\tau - \bar{\mathbf{I}}_{xx}^0(x,t) \ . \tag{24}$$

The latter, when the both sides are derived with respect to t, reduces to

$$u_{tt} = k_0(u_{xx} + 2uu_x) + \int_0^t \left\{ k(t - \tau) \left[e^{D^{-1}[u(x,\tau) - u(x,t)]} [u^2(x, \tau) + u_x(x, \tau)] \right]_x \right\}_t d\tau - \bar{\mathbf{I}}_{xxt}^0(x,t) \ , \tag{25}$$

where $k_0 = k(0)$, $k_0 > 0$. Hence, the initial boundary value problem (2.7) together with the initial and boundary condition (11) - (12), in the 1-dimensional case, is

---

[1] The same property is enjoyed by the generic non homogeneous 1-dimensional equation (13).

[2] No problem in its definition since the 1-dimensional heat conductor is assumed of finite length, however, the infinite length case can be treated in the same way provided the positive valued function v is assumed to belong to the Schwarz's space of functions *rapidly decreasing* together with all their x-derivatives as $x \to \pm\infty$.

related to the nonlinear equation (25) together with the initial and boundary conditions which follow on application of the Cole-Hopf transformation, that is

$$u(x,0) = \frac{v_0'(x)}{v_0(x)}, \quad v_0(x) \neq 0 \tag{26}$$

$$u(x,t)|_{x=0} = u(x,t)|_{x=L}0, \quad \forall t \geq 0 . \tag{27}$$

where the condition of no heat flux through the boundaries is translated into the homogeneous boundary conditions (27) at its two endpoints.

Note that, if (16) is replaced by the Volterra integro-differential equation

$$v_t = \int_{-\infty}^{t} k(t-\tau)v_{xx}(x,\tau)\,d\tau \quad , \quad (x,t) \in [0,L] \times \mathbb{R} \tag{28}$$

the same approach gives the same result; indeed, also the inverse differential operator D_x^{-1} can be defined in the same way.

A slightly different result follows when the thermal conductivity k depends on the space variable too[3]. Accordingly, the evolution equation

$$u_t = \int_0^t \left\{ k(x,t-\tau)e^{D^{-1}[u(x,\tau)-u(x,t)]}[u^2(x,\tau)+u_x(x,\tau)] \right\}_x d\tau - \bar{\mathbf{I}}^0_{xxt}(x,t) , \tag{29}$$

which gives

$$u_{tt} = k(x,0)(u_{xx}+2uu_x) + k_x(x,0)(u^2+u_x)+$$

$$\int_0^t \left\{ k(x,t-\tau)e^{D^{-1}[u(x,\tau)-u(x,t)]}[u^2(x,\tau)+u_x(x,\tau)] \right\}_{xt} d\tau - \bar{\mathbf{I}}^0_{xxt}(x,t) \tag{30}$$

which, again, shows its hyperbolic nature.

However, the applied Cole-Hopf transformation, turns out not to be of help in finding solutions to given initial boundary value problems since the nonlinear equation (25), related to the linear equation (16) which model heat conduction with memory, is not easy to handle.

When the integral term in (25) is omitted, the equation:

$$u_{tt} = k_0(u_{xx}+2uu_x) \tag{31}$$

follows. The latter, not referring to any initial or boundary condition, by inspection, can be checked to admit, further to the trivial solution $u(x,t) = A$, for any arbitrary chosen real constant A, both the traveling wave solutions

$$u_{\pm}(x,t) = \frac{k_0-c^2}{k_0(x \pm ct +A)}, \quad \forall A \in \mathbb{R}, \ \forall c \in \mathbb{R}^+ . \tag{32}$$

---

[3] In such a case the *fading memory* requirement is fulfilled when (3.2) and (3.3) are satisfied at each point of the rigid heat conductor.

In (32), the propagation speed c is arbitrary, however if $c^2 = k_0$, then both solutions reduce to the trivial zero solution. In addition, no matter the value of A, on application of the Cole-Hopf transformation (17), $u_\pm(x,t)$, are transformed into $v(x,t) = B$, where B denotes an arbitrary real constant. A systematic study of equation (31) is currently under investigation; indeed, even if its physical meaning is not clear, as yet, it seems very interesting under the structural viewpoint.

4 Concluding Remarks

The aim of this closing Section is to analyze the obtained result to show perspectives it opens as well as its limits. Finally, connections with other works are given.

First of all, the Cole-Hopf Transformation, here, is applied to a linear evolution equation instead than to a nonlinear one. Indeed, in most of the cases Bäcklund Transformations are applied to obtain a linear equation from a nonlinear one: this is the viewpoint of Rogers and Ruggeri [17], who, considering a nonlinear model of heat conduction, via a reciprocal transformation, related it to a linear telegraph equation. Here, conversely, the idea is to connect a linear integro-differential equation to a hyperbolic nonlinear differential equation, to provide a new justification of the well known hyperbolic behaviour of such a linear integro-differential equation. This viewpoint is currently under investigation to reveal connections with other works as well as, possibly, explicit solutions to some boundary value problems of interest.

Connections between heat conduction and nonlinear hyperbolic type equations have been studied by many authors who were concerned about the physical model: a review on heat waves is due to Joseph and Preziosi [12], subsequent results, also referred to experiments, are given by Saxton and Saxton [18]. However, these models are concerned about some refinements which take into account a richer phenomenology, which originates nonlinearities, with respect to that one of the model here considered.

In addition, the nonlinear equation (25) is not seem to be promising to find new solutions to initial boundary value problems. Conversely, the nonlinear equation (31), even if it provides no solution of the heat conduction problem under investigation, looks interesting in itself under the structural viewpoint. Indeed, it seems it has not been studied as yet. Non linear equations whose higher order part coincides with the second order linear wave operator appear in the work of Lagno et al [13, 14], of Vladimirov and Kutafina [19] and of Vladimirov and Maczka [20] who are concerned about hyperbolic generalization of Burgers equation; however, no choice of the parameters therein comprised produce the nonlinear equation (31); indeed, all the nonlinear equations therein do include various terms with time derivatives.

Acknowledgements. The partial support of G.N.F.M.-I.N.D.A.M. and of PRIN2005 *Mathematical models and methods in continuum physics* are gratefully acknowledged.

References

1. Belleni-Morante, A.: An integro-differential equation arising from the theory of heat conduction in rigid materials with memory. Boll. Unione. Mat. Ital. 15-B(5), 470–482 (1978)
2. Cattaneo, C.: Sulla conduzione del calore. Atti. Sem. Mat. Fis. Universitá Modena 3, 83–101 (1948)
3. Carillo, S.: Existence, Uniqueness & Exponential Decay: an Evolution Problem in Heat Conduction with Memory. Quart. Appl. Math. (2008) (submitted)
4. Cole, J.D.: On a quasilinear parabolic equation occuring in aerodynamics. Quart. App. Math., 9, 225-236 (1951)
5. Coleman, B.D.: Thermodynamics of materials with memory. Arch. Rat. Mech. Anal. 17, 1–46 (1964)
6. Davis, P.L.: On the hyperbolicity of the equations of the linear theory of heat conduction for materials with memory. SIAM J. Appl. Math. 30, 75–80 (1976)
7. Fabrizio, M., Gentili, G., Reynolds, D.W.: On rigid heat conductors with memory. Int. J. Eng. Sci. 36, 765–782 (1998)
8. Giorgi, C., Gentili, G.: Thermodynamic properties and stability for the heat flux equation with linear memory. Quart. Appl. Math. 51(2), 343–362 (1993)
9. Ben-Yu, G., Carillo, S.: Infiltration in soils with prescribed boundary concentration: a Burgers model. Acta Appl. Math. Sinica 6(4), 365–369 (1990)
10. Gurtin, M.E., Pipkin, A.C.: A general theory of heat conduction with finite wave speeds. Arch. Rat. Mech. Anal. 31, 113–126 (1968)
11. Hopf, E.: The partial differential equation $u_t + uu_x = muu_{xx}$. Comm. Pure Appl. Math. 3, 201–230 (1950)
12. Joseph, D.D., Preziosi, L.: Heat waves. Rev. Modern Phys. 61(1), 41–73 (1989); Rev. Modern Phys. 62(2), 375–391 (1990)
13. Lagno, V.I., Zhdanov, R., Magda, O.V.: Group classification and exact solutions of nonlinear wave equations. Acta Appl. Math. 91(3), 253–313 (2006)
14. Lahno, V.I., Zhdanov, R.: Group classification of nonlinear wave equations. J. Math. Phys. 46(5), 053301, 37 (2005)
15. Rogers, C., Shadwick, W.F.: Bäcklund Transformations and their Applications. Mathematics in Science and Engineering, vol. 161. Academic Press, New York (1982)
16. Rogers, C., Ames, W.F.: Nonlinear Boundary Value Problems in Science and Engineering. Academic Press, New York (1989)
17. Rogers, C., Ruggeri, T.: A reciprocal Backlund transformation: application to a nonlinear hyperbolic model in heat conduction. Lett. Nuovo Cimento 44(2, 5), 289–296 (1985)
18. Saxton, K., Saxton, R.: Phase transitions and change of type in low-temperature heat propagation. SIAM J. Appl. Math. 66(5), 1689–1702 (2006) (electronic)
19. Vladimirov, V.A., Kutafina, E.V.: Analytical description of the coherent structures within the hyperbolicgeneralization of burgers equation. Reports on Mathematical Physics 58(3), 465–476 (2007)
20. Vladimirov, V.A., Maczka, C.: Exact Solutions of Generalized Burgers Equation, Describing Travelling Fronts and Their Interaction. Reports on Mathematical Physics 60(2), 317–328 (2007)

Mesoscopic Mechanical Analyses of Textile Composites: Validation with X-Ray Tomography

Pierre Badel, Eric Maire, Emmanuelle Vidal-Sallé, and Philippe Boisse

Abstract. The knowledge of the deformed geometry of woven fabrics is necessary in applications like simulations of permeability of woven fabrics or composite damage simulations. Finite element mesoscopic analyses of textile reinforcements can be used to this aim. The mechanical behavior of the yarns is very specific and the constitutive model used in the simulations is an important aspect of the simulations. The objective of the present paper is to present mesoscopic analyses of woven textile reinforcement deformation based on a specific continuous constitutive model.

The constitutive model has to convey the specificities of the fibrous material of the yarns. The large longitudinal stiffness of the yarns compared to other rigidities and the transverse behavior of the yarns are two crucial aspects for mesoscopic analyses of woven reinforcements. A transversely isotropic model is used within the frame of hypo-elasticity. The objective derivative used in the constitutive equation is specifically defined from the fiber rotation to account for the first aspect. The transverse behavior plays a major role in the final deformed shape of the yarns and the fabric. The identification of the model's parameters is performed via an inverse method.

Large deformation analyses of unit cells (at mesoscopic scale) are presented in order to show the efficiency of the approach. The pure shear behavior of a reinforcement is compared to experimental mechanical tests showing a good agreement.

Keywords: Textile composites, Hypo-elasticity, Fibrous material, In-plane shear.

1 Introduction

The RTM process for composite material forming consists of three stages. A dry textile reinforcement is formed (preforming stage), then the resin is injected within this preform and cured to obtain the final composite part. During the first stage,

Pierre Badel, Emmanuelle Vidal-Sallé, and Philippe Boisse
LaMCoS, CNRS UMR 5259, 20 avenue Einstein, 69621 Villeurbanne, France
Ph.: (33) 4 72 43 61 11; Fax: (33) 4 72 43 85 25
e-mail: Pierre.badel@insa-lyon.fr

Eric Maire
MATEIS, INSA de Lyon, 20 avenue Einstein, 69621 Villeurbanne, France

J.-F. Ganghoffer, F. Pastrone (Eds.): Mech. of Microstru. Solids 2, LNACM 50, pp. 71–78.
springerlink.com © Springer-Verlag Berlin Heidelberg 2010

the reinforcement undergoes in-plane deformations like biaxial extension, in-plane shear, compaction, bending. These deformations can be large especially in-plane shear which is essential in the case of double curved shapes. These macroscopic deformations are directly related to mesoscopic deformations of the reinforcement (at the scale of the yarns). For instance large in-plane shear of the reinforcement leads to a significant lateral crushing of the yarns. As a consequence, the macroscopic behaviour of the fabric is related to its mesoscopic behaviour. Moreover the local deformations can modify the mechanical properties of the reinforcement and its permeability.

The objective of this paper is to present mechanical analyses of a woven composite reinforcement representative unit cell (i.e. at mesoscopic scale). These simulations enable to determine the macroscopic mechanical behaviour at large strain of dry reinforcements. This mechanical behaviour is necessary in finite element simulations of the performing stage. Besides, knowing the deformed geometry of the woven cell enables to determine the permeability of the fibrous reinforcement via Stokes (or Stokes Brinkman) flow simulations within this deformed cell. At last the geometry of the deformed reinforcement heavily influences the mechanical behaviour of the final composite part. In particular, meso-scale damage prediction simulations require knowing this geometry.

The yarn constitutive model used in the following analyses is based on a hypo-elastic approach. The behaviour of the yarn is very specific since it is made of thousands of fibers which can slide with respect to each other. Therefore the objective derivative used in the yarn hypo-elastic constitutive model has to be governed by the fiber direction. The yarn is supposed to be transversely isotropic. The parameters of the material model are identified by an inverse method. In this paper, an example of in-plane shear of a unit cell is shown and compared with experimental results on the mechanical point of view.

2 Yarn Constitutive Behaviour

2.1 Rate Constitutive Equations

Rate constitutive equations (or hypo-elastic laws) are very much used in finite element analyses at large strains [1]. User subroutines that can be implemented in codes such as ABAQUS to define the mechanical constitutive behaviour are written within this framework. A stress rate $\underline{\underline{\sigma}}^{\nabla}$ is related to the strain rate $\underline{\underline{D}}$ by a constitutive tensor $\mathbf{C}$. In order to avoid that rigid body rotations affect the stress state, the derivative $\underline{\underline{\sigma}}^{\nabla}$, called objective derivative, is the derivative for an observer who is fixed with respect to the matter. Because this requirement is not uniquely defined there are several objective derivatives. In this work rotational objective derivatives correspond to a rotation tensor $\underline{\underline{Q}}$ characterizing the rotation of the matter. The rate constitutive equation has the form:

$$\underline{\underline{\sigma}}^{\nabla} = \underline{\underline{\underline{C}}} : \underline{\underline{D}} \text{ with } \underline{\underline{\sigma}}^{\nabla} = \underline{\underline{Q}}.\left(\frac{\mathrm{d}}{\mathrm{dt}}\left(\underline{\underline{Q}}^{\mathrm{T}}.\underline{\underline{\sigma}}.\underline{\underline{Q}}\right)\right).\underline{\underline{Q}}^{\mathrm{T}} = \underline{\underline{\dot{\sigma}}} + \underline{\underline{\sigma}}.\underline{\underline{\Omega}} - \underline{\underline{\Omega}}.\underline{\underline{\sigma}} \qquad (1)$$

where $\underline{\underline{\Omega}}$ is the spin corresponding to $\underline{\underline{Q}}$, i. e. $\underline{\underline{\Omega}} = \underline{\underline{\dot{Q}}}.\underline{\underline{Q}}^{\mathrm{T}}$.

The most usual objective derivatives are those of Green-Naghdi and Jaumann. In the case of the derivative of Green-Naghdi the rotation $\underline{\underline{Q}}$, considered as that of the matter, is the rotation $\underline{\underline{R}}$ of the polar decomposition, which is derived from the decomposition $\underline{\underline{F}}=\underline{\underline{R}}.\underline{\underline{U}}$ of the gradient tensor. In the case of the derivative of Jaumann, $\underline{\underline{Q}}$ is the rotation of the corotational spinless frame, $\underline{\underline{R}}_S$, which is derived from the velocity gradient $\underline{\nabla}\underline{v}$ and the spin $\underline{\underline{\Omega}}_S = \frac{1}{2}\left(\underline{\nabla}\underline{v} - \underline{\nabla}^T\underline{v}\right) = \underline{\underline{\dot{R}}}_S.\underline{\underline{R}}_S^T$.

During a finite element analysis the rate constitutive equation is used to update stresses once the displacement field and the corresponding strain field have been computed over the current time increment. Integrating equation (1) over a time increment $\Delta t = t^{n+1} - t^n$ leads to the widely used formula of Hughes and Winget [2] for stress update:

$$\left[\sigma^{n+1}\right]_{e_i^{n+1}} = \left[\sigma^n\right]_{e_i^n} + \left[C^{n+1/2}\right]_{e_i^{n+1/2}}\left[\Delta\varepsilon\right]_{e_i^{n+1/2}} \text{ with } \left[\Delta\varepsilon\right]_{e_i^{n+1/2}} = \left[C^{n+1/2}\right]_{e_i^{n+1/2}}\left[D^{n+1/2}\right]_{e_i^{n+1/2}}\Delta t \quad (2)$$

where $\left[S\right]_{e_i^n}$ denotes the matrix of the components of the tensor S in the basis $\underline{e}_i \otimes \underline{e}_j \otimes \dots \otimes \underline{e}_m$ at time t^n. Voigt notation is used: the components of a second order tensor are arranged in a single column matrix. The basis of vectors $\underline{e}_i^n$ ($i = 1, 3$) comes from the transportation at time t^n of the initial basis vectors $\underline{e}_i^0$ by the rotation $\underline{\underline{Q}}$ which defines the objective derivative (1). The frame $\{\underline{e}_1^n, \underline{e}_2^n, \underline{e}_3^n\}$ denoted as $\{\underline{e}_i^n\}$ is called rotated frame.

2.2   Textile Composite Reinforcement Mechanical Behaviour

Textile materials are made of fibres, which makes their mechanical behaviour very specific. Relative sliding is possible between fibres (see fig. 1a). The yarns are made of thousands or tens of thousands of fibres and it is in general not possible to model each of them. The constitutive model that is introduced in the present paper is a continuum model intended to stand for the specific mechanical behaviour of the fibre bundle (fig. 1b). In this paper a single fibre direction is considered since the mesoscopic scale is the scale of the yarns. The fiber bundle behavior is supposed to be transversely isotropic. The transverse behaviour is thus assumed to be isotropic (though unhomogeneous). This assumption is supported by high resolution tomography observations [3] made on deformed and undeformed reinforcements (fig. 2).

The equivalent continuum behaviour must take into account the fibrous nature of the material. The fibre direction stiffness is much larger than the others. Consequently the constitutive tensor C is oriented by $\underline{f}_1$ the unit vector in the direction of the fibre. The direction of the vector $\underline{f}_1$ is in general not constant in $\{e_i\}$. Since it is a material direction, the initial fibre direction $\underline{f}_1^0$ is transformed by $\underline{\underline{F}}$, the gradient tensor, into $\underline{f}_1$ while $\{\underline{e}_i\}$ is rotated by $\underline{\underline{Q}}$.

To solve this problem the proposed approach [4] consists in using for equation (1) an objective derivative defined from the fibre rotation.

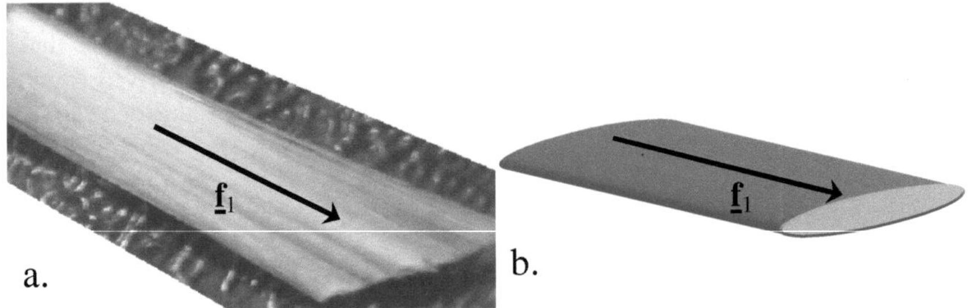

Fig. 1 Fibre direction $\underline{\mathbf{f}}_1$ inside a yarn. (a) actual yarn (b) equivalent continuum model

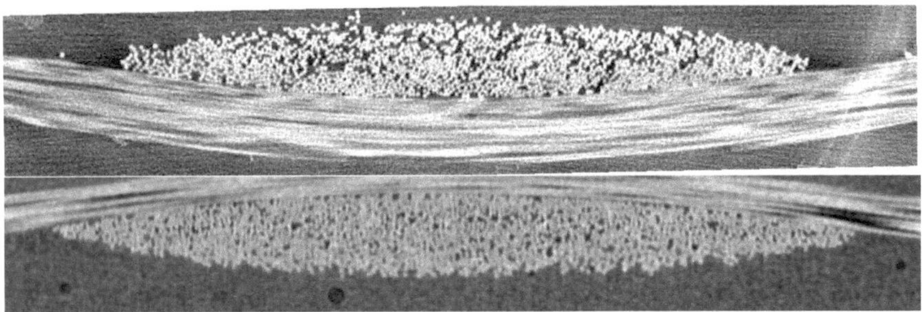

Fig. 2 Tomography reconstructed slices of a glass plain weave. Undeformed state (top) and biaxial tensioned state (bottom). (different scales)

2.3 *Objective Derivative Based on the Fibre Rotation*

In this approach, the rate constitutive equation (1) is based on the rotation of the fibre $\underline{\mathbf{f}}_1$:

$$\underline{\underline{\boldsymbol{\sigma}}}^{\nabla_\phi} = \underline{\underline{\mathbf{C}}} : \underline{\underline{\mathbf{D}}} \quad \text{with} \quad \underline{\underline{\boldsymbol{\sigma}}}^{\nabla_\phi} = \underline{\underline{\boldsymbol{\Phi}}}.\left(\frac{\mathrm{d}}{\mathrm{dt}}\left(\underline{\underline{\boldsymbol{\Phi}}}^{\mathrm{T}}.\underline{\underline{\boldsymbol{\sigma}}}.\underline{\underline{\boldsymbol{\Phi}}} \right) \right).\underline{\underline{\boldsymbol{\Phi}}}^{\mathrm{T}} \tag{3}$$

where $\underline{\underline{\boldsymbol{\Phi}}}$ is the rotation of the fibre. It can be shown that this derivative is objective. The stress update (2) becomes:

$$\left[\boldsymbol{\sigma}^{n+1} \right]_{f_i^{n+1}} = \left[\boldsymbol{\sigma}^n \right]_{f_i^n} + \left[\mathbf{C}^{n+1/2} \right]_{f_i^{n+1/2}} \left[\Delta\boldsymbol{\varepsilon} \right]_{f_i^{n+1/2}} \tag{4}$$

The rotation $\underline{\underline{\boldsymbol{\Phi}}}$ (eq. (3)) from the initial known frame $\{\underline{\mathbf{f}}_i^0\}$ to the current frame $\{\underline{\mathbf{f}}_i\}$ has to be determined. From the transformation gradient $\underline{\underline{\mathbf{F}}}$, the current fibre direction $\underline{\mathbf{f}}_1$ can be determined. Assuming that the initial position of the fibre is $\underline{\mathbf{f}}_1^0$:

$$\underline{\mathbf{f}}_1 = \frac{\underline{\underline{\mathbf{F}}} \cdot \underline{\mathbf{f}}_1^0}{\left\| \underline{\underline{\mathbf{F}}} \cdot \underline{\mathbf{f}}_1^0 \right\|}$$

The other basis vectors $\underline{\mathbf{f}}_2$ and $\underline{\mathbf{f}}_3$ of the orthonormal frame $\{\underline{\mathbf{f}}_i\}$ are obtained from the material transformation of $\underline{\mathbf{f}}_2^{\,0}$:

$$\underline{\mathbf{f}}_2 = \frac{\underline{\underline{\mathbf{F}}}\cdot\underline{\mathbf{f}}_2^{\,0} - \left(\underline{\underline{\mathbf{F}}}\cdot\underline{\mathbf{f}}_2^{\,0}\cdot\underline{\mathbf{f}}_1\right)}{\left\|\underline{\underline{\mathbf{F}}}\cdot\underline{\mathbf{f}}_2^{\,0} - \left(\underline{\underline{\mathbf{F}}}\cdot\underline{\mathbf{f}}_2^{\,0}\cdot\underline{\mathbf{f}}_1\right)\right\|} \quad \text{and} \quad \underline{\mathbf{f}}_3 = \underline{\mathbf{f}}_1 \times \underline{\mathbf{f}}_2$$

Then the rotation $\underline{\underline{\boldsymbol{\Phi}}}$ is derived in the following way:

$$\underline{\underline{\boldsymbol{\Phi}}} = \underline{\mathbf{f}}_i \otimes \underline{\mathbf{f}}_i^{\,0} = \left(\underline{\mathbf{f}}_j\right)_i^0 \underline{\mathbf{f}}_i^{\,0} \otimes \underline{\mathbf{f}}_j^{\,0} = \left(\underline{\mathbf{f}}_j\cdot\underline{\mathbf{f}}_i^{\,0}\right)\underline{\mathbf{f}}_i^{\,0} \otimes \underline{\mathbf{f}}_j^{\,0}$$

The main interest of this approach is that the constitutive matrix in equation (4) appears in the frame of the fibre and consequently it is directly in its specific form corresponding to the textile material under consideration. This constitutive matrix written in the fibre frame can be assumed constant in some cases. Generally it is not; the transverse behaviour of a fibrous yarn is depending on the strain state.

When using a material user subroutine in a code such as ABAQUS, the strain increment $\Delta\varepsilon$ is given at Gauss points in a frame which is not $\{\underline{\mathbf{f}}_i\}$ but a standard frame. In the case of ABAQUS/Explicit it is Green-Naghdi's frame $\{\underline{\mathbf{e}}_i\}$ ($\underline{\mathbf{e}}_i = \underline{\underline{\mathbf{R}}}\cdot\underline{\mathbf{e}}_i^{\,0}$). To use equation (4) it is necessary to calculate $\left[\Delta\varepsilon\right]_{f_i}$ by a change of basis corresponding to the rotation $\underline{\underline{\boldsymbol{\Phi}}}.\underline{\underline{\mathbf{R}}}^T$. In the same way, when the stress update is performed with equation (4) it is necessary to return the stress components at t^{n+1} in the code's work frame using an inverse change of basis (corresponding to $\underline{\underline{\mathbf{R}}}.\underline{\underline{\boldsymbol{\Phi}}}^T$).

2.4   Constitutive Matrix

Thanks to the use of the fiber frame, the constitutive matrix components along the fibre direction and the transverse ones can be distinguished.

The fibre direction modulus is obtained by a tensile test on a yarn. It is considered as constant. The transverse modulus (remember that the transverse behaviour is assumed to be isotropic) is related to the longitudinal and transverse strains. If the yarn undergoes longitudinal tension, it becomes transversely much stiffer. It is also very little stiff transversally when transverse compression is low and it becomes stiffer as compression increases. The following form is used for the transverse modulus E_T:

$$E_T(c, \varepsilon_{11}) = E_0 + k|\varepsilon_{11}|c^2$$

where c is a measure of the transverse compaction namely the local cross section area variation and ε_{11} is the longitudinal strain. The coefficient values E_0 and k have been identified by an inverse method from equi-biaxial tension tests which lead to a significant compaction of the yarn [5]. It was shown in [5] that in order to have a continuum with a yarn type behaviour, i.e. null bending stiffness (or very low), as it is the case for a bundle of several thousands of fibres because of relative sliding of fibres, shear moduli have to be null or very low. Poisson ratios are supposed to be null. The values of the material properties used for the glass plain weave of this study are listed in Table 1.

Table 1. Material parameters

Longitudinal Young modulus E1	35400 MPa
Transverse Young modulus	$0.2 + 8.10^4|\varepsilon_{11}|c^2$ MPa
Poisson ratios	0
Shear moduli	20 MPa

3 Mesoscopic Simulation of the Shear of a a Glass Plain Weave Unit Cell

The hypo-elastic material model based on the fibre rotation, introduced in section 2, is used to simulate the in-plane shear of a woven unit cell. There are two types of objectives for such a simulation. Firstly the macroscopic shear mechanical behaviour of the reinforcement can be determined, which is difficult experimentally. The picture frame and bias tests that are used to this aim are delicate. The mesoscopic simulation can also be performed at the design stage of the reinforcement. Secondly, it provides local mesoscopic results such as yarn deformation and shape. These results are very important to perform damage prediction analyses or to determine permeabilities of the reinforcement.

The fabric studied in this paper is a glass plain weave of which the specifications are given in Table 2. The geometric model is established in order to insure its consistency, i.e. there are neither yarn penetrations nor unexpected voids [6]. The mesh of the yarns is mapped, which allows easily defining the initial fibre direction. Next, the choice of the unit cell and boundary conditions has to render the periodicity of the reinforcement. To this end the displacement field is split into a macroscopic average part and a local periodic part, which easily leads to defining kinematical boundary conditions for each material point of the boundary. More details about other requirements can be found in [7].

The material parameters (see Table 1) are determined from a tensile test on a single yarn for the Young modulus and from an inverse method with an equi-biaxial tension test for the other parameters. The latter method is interesting to determine the transverse behaviour because it features significant transverse crushing of warp yarns over weft yarns. At last, the friction coefficient it set to 0.2, which is a usual value for friction between carbon fibres.

Table 2. Balanced glass plain weave specifications

Weaving	Plain weave
Yarn width (mm)	Warp: 3,2 Weft: 3,1
Densities (Yarn/mm)	Warp: 0,251 Weft: 0,248
Crimp (%)	Warp: 0,5 Weft: 0,54
Surface weight (g/m²)	600

Figure 3a shows the shear curve extracted from the computations compared with the one of an experimental picture frame test. From 35° the stiffness increases significantly with the shear angle because of the yarn locking within the woven cell. From this transition, the shear stiffness is related to lateral crushing of the yarns due to the square geometry turning into a rhomboid one. The agreement is good, keeping in mind that spurious tensions are very difficult to avoid in the picture frame test and tend to slightly overestimate the shear curve at large shear angles. This is the main reason why results may be different from one test to another.

The deformed cell for a shear angle of 53° is shown in fig. 3b. The local compaction is plotted and this value can locally reach 39%. Though this deformed geometry doesn't suffer any major distortion or defects, it can not be evaluated further at the moment. It is planed to use tomography as an observation tool of deformed woven reinforcements in order to compare the obtained images with the simulated shape.

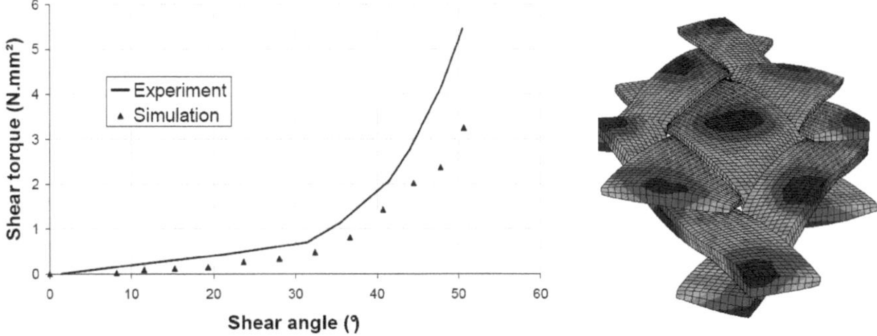

Fig. 3 (a) shear curve, simulation vs experiment (b) deformed unit cell with contours of local compaction

4 Conclusion

A method for the analysis of mesoscopic deformation of woven reinforcements was presented. A very important aspect is the yarn constitutive model which is based on a specific transversely isotropic hypo-elastic model developed for large strain analysis of textile composite reinforcements. It is based on an objective derivative defined from the fibre rotation. The approach is simple and can be implemented in any commercial F.E. software. Nevertheless it must be underlined that its efficiency will depend on the quality of the identification of the constitutive matrix [C]. Especially the transverse moduli mainly depend on the fibre compaction (i.e. on strains) and their values are very important for the accuracy of the simulation.

From these analyses, the mechanical behaviour of the reinforcement can be determined. It is used in several applications like finite element forming simulations or at the design stage of a fabric. The determination of the deformed geometry of the reinforcement is another important result since it is useful in fluid flow simulations aimed at computing the permeability. It can also be used for damage prediction simulation.

The results presented are good compared to the experiments. However, further investigations are needed to evaluate the deformed geometry. This is the point of the current work in progress using tomography as an observation tool.

References

1. Criesfield, M.A.: Non linear Finite Element Analysis of Solids and Structure: Advanced Topics, vol. 2. John Wiley, Chichester (1997)
2. Hughes, T.J.R., Winget, J.: Finite rotation effects in numerical integration of rate constitutive equations arising in large deformation analysis. Int. J. Num. Meth. Eng. 15, 1862–1867 (1980)
3. Baruchel, J., Buffiere, J.Y., Maire, E., Merle, P., Peix, G.: X ray tomography in material science, Hermès eds, Paris (2000)
4. Hagège, B., Boisse, P., Billoët, J.L.: Finite element analyses of knitted composite reinforcement at large strain. Eur. J. Comput. Mech. 14, 767–776 (2005)
5. Gasser, A., Boisse, P., Hanklar, S.: Mechanical behaviour of fry fabric reinforcements. 3D simulations versus biaxial test, Comp. Mat. Sci. 17, 7–20 (2000)
6. Hivet, G., Boisse, P.: Consistent 3D geometrical model of fabric elementary cell. Application to a meshing preprocessor for 3D finite element analysis, Finite Elem. Anal. Des. 42, 25–49 (2005)
7. Badel, P., Vidal-Sallé, E., Boisse, P.: Computational determination of in-plane shear mechanical behaviour of textile composite reinforcements. Comput. Mater. Sci. 40, 439–448 (2007)

Mechanical Response of Helically Wound Fiber-Reinforced Incompressible Non-linearly Elastic Pipes

Paola Nardinocchi, Tomas Svaton, and Luciano Teresi

Abstract. We study the mechanical response of a helically wound fiber–reinforced incompressible axisymmetric structure under torsion and compare it with the response turning out from the classical Rivlin solution of the torsion problem of a neo-Hookean pipe.

1 Introduction

In the last few years, fiber–reinforced materials have made a new comeback due to the bio–mechanical involvement. Indeed, a lot of biological tissues are characterized by a material response that is strongly anisotropic due to a fiber structure which determines the relevant mechanical behavior of the tissue and the problem of characterizing the behavior of fiber–reinforced media undergoing finite deformations turns out as a relevant theme in literature [1], [2]. The macroscopic description of the material response of fiber–reinforced materials is typically given in terms of a strain-energy function which is dependent on specific deformation invariants [8]. Moreover, it is usually assumed that such function be additively decomposable into an isotropic part associated with the isotropic base material and an anisotropic part accounting for the anisotropic character of the material due to the reinforcement. Recently, a lot of analysis has been devoted to show as different phenomena related to fiber–reinforced materials may be captured within this framework [3], [4], [5].

Paola Nardinocchi
Dipartimento di Ingegneria Strutturale e Geotecnica, Università di Roma "La Sapienza"
via Eudossiana 18, I-00147 Roma, Italy
e-mail: `paola.nardinocchi@uniroma1.it`

Tomas Svaton
University of West Bohemia, Pilsen, Czech Republic
e-mail: `tsvaton@gmail.com`

Luciano Teresi
Dipartimento di Strutture, Università Roma Tre, Roma, Italy
e-mail: `teresi@uniroma3.it`

J.-F. Ganghoffer, F. Pastrone (Eds.): Mech. of Microstru. Solids 2, LNACM 50, pp. 79–87.
springerlink.com
© Springer-Verlag Berlin Heidelberg 2010

Here, we analyze the mechanical response of a helically wound fiber–reinforced incompressible cylindrical structure to torsional deformations. The pure torsional deformation has been studied for isotropic and incompressible cylinders by Rivlin [7]. We extend the Rivlin's solution to helically wound fiber–reinforced incompressible cylinders and cylindrical pipes and study the rich mechanical behaviour due to the anisotropy induced by the fibers. From our point of view, this is a first step towards the analysis of the torsional deformation induced during the cardiac cycle in the left ventricles which may be roughly viewed as an ellipsoid of revolution consisting of a non–linearly elastic and incompressible material with embedded helicoidal muscle fibers whose angle changes linearly across the wall (recent studies have shown that the left ventricular torsion is an important indicator of cardiac function).

2   Fiber-Reinforced Bodies and Torsional Deformations

In 1940s, R.S. Rivlin wrote a short series of papers on the large elastic deformations of isotropic and incompressible materials with special attention to some simple problems: flexure, shear, and torsion of cylinder-like axisymmetric bodies. Here, the Rivlin solution of the pure torsion problem of elastic, incompressible and cylinder-like bodies is extended to helically wound fiber-reinforced incompressible elastic pipes and cylinders. The classical Rivlin solution is recovered as a special case.

Let $\{\mathbf{e}_1, \mathbf{e}_2, \mathbf{e}_3\}$ be an orthonormal basis of the vector space $\mathscr{V} = T\mathscr{E}$, with $\mathscr{E}$ the three–dimensional Euclidean space, and let $\mathscr{U}$ be the orthogonal complement to span$\{\mathbf{e}_3\}$ with respect to $\mathscr{V}$. In a plane $\mathscr{P} \perp \mathbf{e}_3$ let us consider the connected and compact domain $\mathscr{D} = \mathscr{D}_o \setminus \Lambda_o$ with boundary $\partial \mathscr{D} = \partial \mathscr{D}_o \bigcup \partial \Lambda_o$ where $\mathscr{D}_o$ and the lacuna Λ_o, if present, are simply connected regions. Fixed an interval $\mathscr{I} = [0, h] \subset \mathscr{R}$, we consider the axisymmetric cylinder-like region $\mathscr{C} = \mathscr{D} \times \mathscr{I}$; $\mathscr{D} \times \{0\}$ and $\mathscr{D} \times \{h\}$ are the bases of $\mathscr{C}$, and $\partial \mathscr{D} \times \mathscr{I}$ is the mantle of $\mathscr{C}$, eventually given by the inner mantle $\partial \Lambda_o \times \mathscr{I}$ and the outer mantle $\partial \mathscr{D}_o \times \mathscr{I}$. Fixed a pole $o \in \mathscr{P}$, the place $\mathbf{p}$ of any $p \in \mathscr{C}$ with respect to o is given by the vector field

$$\mathbf{p} = r\mathbf{n}(\theta) + \zeta \mathbf{e}_3, \quad \mathbf{n}(\theta) = \cos\theta\, \mathbf{e}_1 + \sin\theta\, \mathbf{e}_2, \, 0 < \theta < 2\pi, \, \zeta \in \mathscr{I}; \quad (1)$$

moreover, $0 < r < R_o$ when $\mathscr{D} \equiv \mathscr{D}_o$, $R_i < r < R_o$ otherwise[1]. A helix-like fiber of pitch b is a curve whose unit tangent vector $\mathbf{e} = \hat{\mathbf{e}}(r, \theta)$ at any $p \in \mathscr{C}$ is defined as

$$\hat{\mathbf{e}}(r, \theta) = \frac{r}{(r^2 + b^2)^{\frac{1}{2}}} \mathbf{n}_{,\theta}(\theta) + \frac{b}{(r^2 + b^2)^{\frac{1}{2}}} \mathbf{e}_3. \quad (2)$$

So, at any place $\mathbf{p}$ of a helically wound fiber–reinforced body $\mathscr{C}$, the material fiber $(\mathbf{p}, \mathbf{e})$ with $\mathbf{e}$ as in (2) identifies a locus of material reinforcement; for $b = 0$ equation (2) defines a circumferential uniaxial reinforcement. An interesting case is represented by a helically wound fiber–reinforced cylindrical pipe with pitch $b = b(r)$.

---

[1] With this, an element $p \in \mathscr{C}$ is equivalently identified by the place $\mathbf{p}$ and by the cylindrical coordinates (r, θ, ζ).

2.1 The Pure Torsional Deformation: Strain and Stress

In [7], Rivlin introduces the torsional deformation $\mathscr{C} \ni p \mapsto x = \mathbf{f}(p) \in \mathscr{E}$ of the cylinder–like body in which planes orthogonal to $\mathrm{span}\{\mathbf{e}_3\}$ remain plane and suffer only a pure rotation about $\mathbf{e}_3$. A concise representation formula for $\mathbf{f}$ is the following

$$\mathbf{f}(\mathbf{p}) = (\mathbf{I} + \sin\varphi\, \mathbf{e}_2 \wedge \mathbf{e}_1 - (1 - \cos\varphi)\,\check{\mathbf{I}})\,\mathbf{p}, \quad \varphi = \zeta\,\tau, \tag{3}$$

where $\mathbf{e}_2 \wedge \mathbf{e}_1 = \mathbf{e}_2 \otimes \mathbf{e}_1 - \mathbf{e}_1 \otimes \mathbf{e}_2$, φ is the *torsion angle*, and τ is the unit torsion angle. From now on, we parametrize the torsional deformation (3) through the torsion angle $\hat{\varphi} = h\tau$ of the basis $\mathscr{D} \times \{h\}$. The geometrical structure of the cylinder–like body $\mathscr{C}$ is retained under the deformation (3) and, denoted with $\mathbf{x} \in \mathscr{V}$ the place of x with respect to o, it holds:

$$\mathbf{x} = \mathbf{x}_o + \zeta\,\mathbf{e}_3, \quad \mathbf{x}_o = r(\cos\varphi\,\mathbf{n}(\theta) + \sin\varphi\,\mathbf{n}_{,\theta}(\theta)). \tag{4}$$

We note that, $(\mathbf{x}_o \cdot \mathbf{x}_o)^{1/2} = r$, that is, as expected, the deformation (3) preserves the radius of the cylinder-like body. The deformation gradient $\mathbf{F}$ corresponding to (3) may be represented as

$$\mathbf{F} = \check{\mathbf{F}} + \tau\mathbf{x}_o^* \otimes \mathbf{e}_3 + \mathbf{e}_3 \otimes \mathbf{e}_3, \tag{5}$$

with $\check{\mathbf{F}} \in \mathbb{L}\mathrm{in}(\mathscr{U})$ and $\mathbf{x}_o^* \in \mathscr{U}$ given by

$$\check{\mathbf{F}} = \cos\varphi\,\check{\mathbf{I}} + \sin\varphi\,(\mathbf{e}_2 \wedge \mathbf{e}_1), \quad \mathbf{x}_o^* = \mathbf{e}_3 \times \mathbf{x}_o. \tag{6}$$

From equations (5) and (6) it is easy to verify that the deformation (3) identically satisfies the incompressibility condition $J = \det\mathbf{F} = 1$. We describe the material response of the body through the strain energy function

$$\psi = c_1(I_1 - 3) + c_2(I_2 - 3) + \frac{1}{2}c_1\gamma(I_4 - 1)^2, \quad J = 1, \tag{7}$$

with I_1 and I_2 the linear and quadratic invariant of the Cauchy–Green tensor $\mathbf{C} = \mathbf{F}^T\mathbf{F}$ corresponding to $\mathbf{F}$ and $I_4 = \mathbf{Ce} \cdot \mathbf{e}$. The strain energy (7) describes the mechanical response of an isotropic and incompressible base material with uniaxial reinforcement (see [3]). The corresponding Cauchy stress is

$$\mathbf{T} = 2(\psi_1 + I_1\,\psi_2)\mathbf{B} - 2\psi_2\mathbf{B}^2 + 2\bar{\psi}_4(I_4 - 1)\mathbf{Fe} \otimes \mathbf{Fe} - p\mathbf{I}, \tag{8}$$

with $\psi_\alpha = \frac{\partial\psi}{\partial I_\alpha} = c_\alpha$, $(\alpha = 1, 2)$, $\mathbf{B} = \mathbf{FF}^T$, and $\bar{\psi}_4 = c_1\gamma$; the Cauchy stress corresponding to an isotropic and incompressible material turns out by setting $\bar{\psi}_4 = 0$ (corresponding to $\gamma = 0$). We have

$$I_1 = 3 + \tau^2 r^2, \qquad I_4 = 1 + \frac{r^2}{b^2 + r^2}(2 + b\tau)\,b\tau. \tag{9}$$

The behavior of I_4 is discussed in figure 1. As first, we note on the left that for circumferential fibers (b/h=0) $I_4 \equiv 1$, that is, the fibers length does not change under

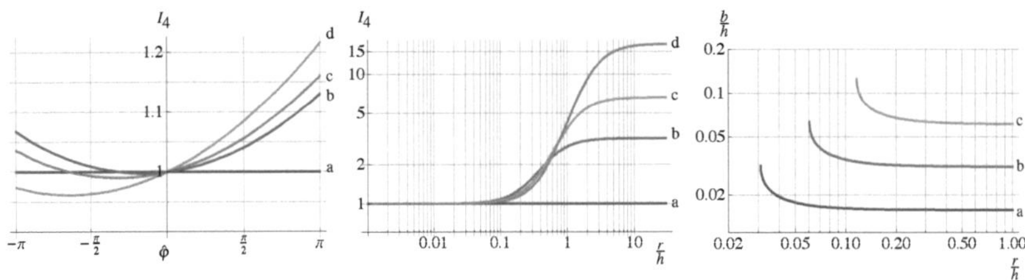

Fig. 1 I_4 VS $\hat{\varphi}$ at r/h=0.1 (left) and I_4 VS r/h at $\hat{\varphi} = \pi/2$ (centre, log-log plot) for four different pitch/height ratios: b/h=0 (a), 1/2 (b), 1 (c), 2 (d). b/h VS r/h (right, log-log plot) to achieve a constant fiber strain along the radius in a torsion with $\hat{\varphi} = \pi/2$: $I_4 = 1.05$ (a), 1.10 (b), 1.20 (c)

the torsional deformation (3); otherwise, in a counterclockwise torsional rotation the stretch of the fibers is monotone increasing with b/h. The log-log plot in the center of Fig. 1 shows the dependence of I_4 on r/h for a given torsional angle $\hat{\varphi} = \pi/2$: for the pitch/height ratios considered, the fibers' stretch is quite low for $r << h$ and, interestingly, has a large plateau for $r >> h$. It is worth noting the possibility of winding the fibers around the cylinder with a variable pitch so that, for a given torsion, the stretch remains constant along the radius; right plot in Fig. 1 shows pitch rate b/h versus r/h that realize three different constant stretches in the fiber for $\hat{\varphi} = \pi/2$.

2.2 The Balance Equations and the Stress Field

The peculiar geometric structure of the body suggests to using the additive decomposition introduced in the representation formula of the deformation gradient $\mathbf{F}$ and writing the Cauchy stress as

$$\mathbf{T} = \check{\mathbf{T}} + 2\mathrm{sym}\,(\mathbf{t} \otimes \mathbf{e}_3) + \sigma \mathbf{e}_3 \otimes \mathbf{e}_3 , \tag{10}$$

in terms of the plane stress component $\check{\mathbf{T}} \in \mathbb{L}\mathrm{in}(\mathscr{U})$, the plane tangential stress vector $\mathbf{t} \in \mathscr{U}$, and the axial stress component $\sigma \in \mathscr{R}$. The standard balance equations of mechanics may be decomposed in agreement with the geometrical structure of $\mathscr{C}$ into a plane vector component and an axial scalar component as

$$\mathrm{div}\check{\mathbf{T}} + \mathbf{t}' = \mathbf{0} \quad \text{and} \quad \sigma' + \mathrm{div}\mathbf{t} = 0, \tag{11}$$

respectively, where div is the divergence operator in $\mathscr{U}$ and a prime denotes the derivative with respect to ζ. When $\mathbf{T}$ is given by the equation (8), we find:

$$\check{\mathbf{T}} = (-p + 2(\psi_1 + \psi_2(I_1 - 1)))\check{\mathbf{I}}$$
$$+ (2\tau^2\psi_1 + 2\bar{\psi}_4(I_4 - 1)\frac{(1 + b\tau)^2}{(b^2 + r^2)})\mathbf{x}_o^* \otimes \mathbf{x}_o^* , \tag{12}$$

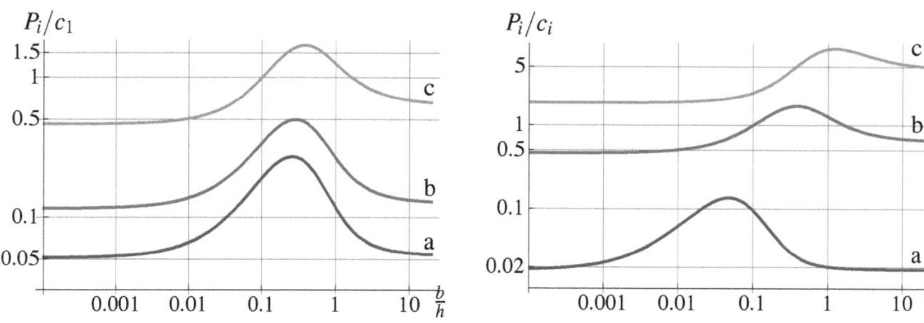

Fig. 2 Inflating pressure VS b/h for three different torsion angles: $\hat{\varphi} = \pi/6$ (a), $\pi/4$ (b), $\pi/2$ (c), with $R_o/h = 0.5$, $R_i/R_o = 0.5$, $\gamma = 1$ (left). Inflating pressure VS b/h for three different aspect ratios of the pipe: $R_o/h = 0.1$ (a), 0.5 (b), 1 (c), with $R_i/R_o = 0.5$, $\gamma = 1$, $\hat{\varphi} = \pi/2$ (right)

$$\mathbf{t} = 2((\psi_1 + \psi_2)\tau + 2\bar{\psi}_4(I_4 - 1)b\frac{(1+b\tau)}{(b^2 + r^2)})\mathbf{x}_o^*, \tag{13}$$

$$\sigma = -p + 2(\psi_1 + 2\psi_2) + 2\bar{\psi}_4(I_4 - 1)\frac{b^2}{(b^2 + r^2)}. \tag{14}$$

So, it is easy to check from equation (13) that $\mathrm{div}\,\mathbf{t} = 0$ and $\mathbf{t}' = \mathbf{0}$; moreover, equation (14) shows that $\sigma' = p'$ and equations (11) yield

$$\mathrm{grad}\,p = -(2(\psi_1 - 2\psi_2)\tau^2 + 2\bar{\psi}_4 b\tau(1+b\tau)^2(2+b\tau)\frac{r^2}{(b^2 + r^2)^2})\mathbf{x}_o, \tag{15}$$

$$-p' = 0. \tag{16}$$

With this, we have that $p = \hat{p}(r)$ and

$$\hat{p}(r) = p_o - (\psi_1 - 2\psi_2)\tau^2 r^2$$
$$- \bar{\psi}_4 b\tau(1+b\tau)^2(2+b\tau)(\frac{b^2}{b^2+r^2} + \log(b^2 + r^2)), \tag{17}$$

with p_o determined by the boundary conditions. One of the goals of the work of R.S. Rivlin ([7]) was to find which kind of tractions, if any, should be applied on the boundary of the cylinder-like region $\mathscr{C}$ which were consistent with the torsional deformation (3). He found that a no–traction condition on $\partial\mathscr{D}_o \times \mathscr{I}$ is consistent with a pure torsional deformation; but, a pressure field $P_i\mathbf{m}$ has to be applied on $\partial\Lambda_o \times \mathscr{I}$ if the lacuna Λ_o is present. For a helically wound fiber–reinforced cylinder–like region, we find analogous results. In absence of the lacuna Λ_o, the no–traction condition $\mathbf{Tm} = \mathbf{0}$ on $\partial\mathscr{D}_o \times \mathscr{I}$ may be written as $\check{\mathbf{T}}\mathbf{m} = \mathbf{0}$ on $\partial\mathscr{D}_o \times \mathscr{I}$; then, equation (17) turns out

$$p_o = 2(\psi_1 + 2\psi_2) + \tau^2 R_o^2 \psi_1$$
$$+ \bar{\psi}_4 b\tau(2+b\tau)(1+b\tau)^2(\frac{b^2}{b^2+R_o^2} + \log(b^2 + R_o^2)); \tag{18}$$

and, with this, the Cauchy stress is completely determined. In presence of the lacuna Λ_o, the no–traction condition on $\partial \mathscr{D}_o \times \mathscr{I}$ is still consistent with the pure torsional deformation (3) and determines completely the Cauchy stress. Thus, it happens that on the boundary $\partial \Lambda_o \times \mathscr{I}$ there is a traction field $\mathbf{Tm} = -P_i \mathbf{m}$ different from zero with

$$P_i = \psi_1 (R_o^2 - R_i^2)\left(\frac{\hat{\varphi}}{h}\right)^2$$
$$+ \bar{\psi}_4 \left(b^2 \frac{R_i^2 - R_o^2}{(b^2 + R_i^2)(b^2 + R_o^2)} + \log \frac{b^2 + R_o^2}{b^2 + R_i^2} \right) \frac{b}{h} \hat{\varphi} \left(1 + \frac{b}{h}\hat{\varphi}\right)^2 \left(2 + \frac{b}{h}\hat{\varphi}\right). \quad (19)$$

It means that a pure torsional deformation can be maintained on a cylindrical pipe iff on the inner mantle a pressure constant field $P_i = \hat{P}_i(R_i, R_o, b, \tau, c_1, \gamma)$ is applied. The pressure field P_i is a fourth order polynomial function of $\hat{\varphi}$, is a non monotone function of the pitch ratio b/h, and contains a stiffness term depending on the geometric characteristics of the pipe R_o/h, R_i/h. Figure 2 summarizes the interesting features of P_i; the non dimensional ratio P_i/c_1 versus b/h is represented in a log-log plot for three different torsion angles (left) and for three different aspect ratios of the pipe (right). On the bases of the cylindrical structure, we find a traction field $\mathbf{T}_\pm \mathbf{e}_3 = \sigma_\pm \mathbf{e}_3 + \mathbf{t}_\pm$ with $\sigma_\pm$ and $\mathbf{t}_\pm$ denoting the values attained by the corresponding fields σ, and $\mathbf{t}$ on the bases $\mathscr{D} \times \{h\}$ and $\mathscr{D} \times \{0\}$, respectively (and $\mathbf{T}_\pm$ is the corresponding Cauchy stress field on the bases)[2]. The Rivlin's result may be recovered by setting $\bar{\psi}_4 = 0$ in the equations (17) and (19). Moreover, we note that for $b = 0$ (circumferential fibers) the stress $\mathbf{T}$ due to a pure torsional deformation does not depend on $\bar{\psi}_4$. It means that circumferential fibers does not reinforce the cylindrical structure with respect to a torsional deformation.

3 Stress Resultants and Torsion Tests

We measure both the torque due to the tractions $\mathbf{t}_\pm$ inducing the torsional deformation and the axial force due to the restrained axial displacement of the structure. The aim is the evaluation of the dependence of the torque $\mathbf{M} = M\mathbf{e}_3$ and the axial force $\mathbf{N} = N\mathbf{e}_3$ on the torsion angle $\hat{\varphi}$, on the modulus γ and on the pitch b of the helix for different cylinder–like structures. The torque and the axial force are determined by the scalar fields

$$M = \int_{\mathscr{P}} \mathbf{x}_o \times \mathbf{t}\, dx_o \cdot \mathbf{e}_3, \quad \text{and} \quad N = \int_{\mathscr{P}} \sigma\, dx_o. \quad (20)$$

---

[2] Their role is well different: the axial stress $\sigma_\pm$ are reactive stresses corresponding to the kinematical constraint on the axial displacement; the tangential stress $\mathbf{t}_\pm$ represent the stress fields to be applied on the bases with the aim to produce a pure torsional deformation.

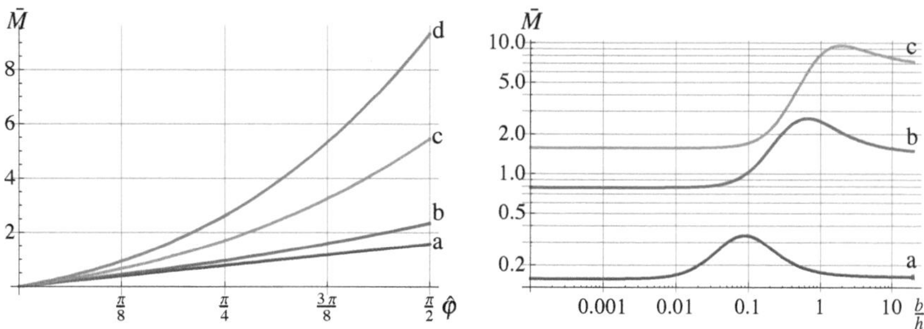

Fig. 3 Torque VS $\hat{\varphi}$ for different fiber stiffness: $\gamma = 0$ (a), 0.1 (b), 0.5 (c), 1 (b) (left); Torque VS b/h for different aspect ratios of the cylinder: Ro/h=0.1 (a), 0.5 (b), 1 (c) (right)

By using equations $(4)_2$ and (13), we find

$$M = \pi R_o^3 c_1 \left(\frac{R_o}{h} \hat{\varphi} + 8\gamma \frac{R_o}{h} f(R_o,b)\, \hat{\varphi} \left(1 + \frac{b}{h} \hat{\varphi}\right) \left(2 + \frac{b}{h} \hat{\varphi}\right) \right), \qquad (21)$$

with

$$f(R_o,b) = \frac{b^2}{R_o^2} \log \frac{b^2}{b^2 + R_o^2} + \frac{R_o^2 + 2b^2}{2(b^2 + R_o^2)}. \qquad (22)$$

In (21), the first addendum is the torque competing to an isotropic and incompressible cylinder, a linear function of $\hat{\varphi}$, corresponding to the Rivlin solution; the second addendum defines the correction due to the presence of the helicoidal fibers. The dependence of the dimensionless torque $\bar{M} = M/\pi R_o^3 c_1$ on the relevant parameters is shown in figure 3: $\bar{M}(\hat{\varphi})$ is an increasing monotone function with respect to $\hat{\varphi}$ and γ (left); more interesting, for any given $\hat{\varphi}$, $\bar{M}$ is not monotone with b/h, and the maximum depends on R_o/h, a parameter measuring the slenderness of the cylinder (right). As far as the axial force N is concerned, we find

$$N = -\frac{1}{2}\pi R_o^2 c_1 \left(\frac{R_o}{h}\right)^2 \hat{\varphi}^2$$

$$+ 2\pi R_o^2 c_1\, \gamma \hat{\varphi} \left(2 + \frac{b}{h} \hat{\varphi}\right) \left(g_1(R_o,b) - \frac{1}{2}\left(1 + \frac{b}{h} \hat{\varphi}\right)^2 f_1(R_o,b) \right) \qquad (23)$$

with

$$f_1(R_o,b) = 2f(R_o,b), \quad g_1(R_o,b) = \frac{R_o^2}{2(b^2 + R_o^2)} - f(R_o,b).$$

The first row of equation (23) defines the axial force competing to the isotropic and incompressible cylinder; as it is expected, in the linear approximation (that is, for small $\hat{\varphi}$) this term goes to zero. The second row gives the contribution to the axial force due to the anisotropy; it is worth noting that such addendum is different from zero in a linear theory. Figure 4 shows the dependence of the dimensionless axial force $\bar{N} = N/\pi R_o^2 c_1$ on $\hat{\varphi}$ for different b/h, and for cylinders with $R_o/h = 0.5$ (left),

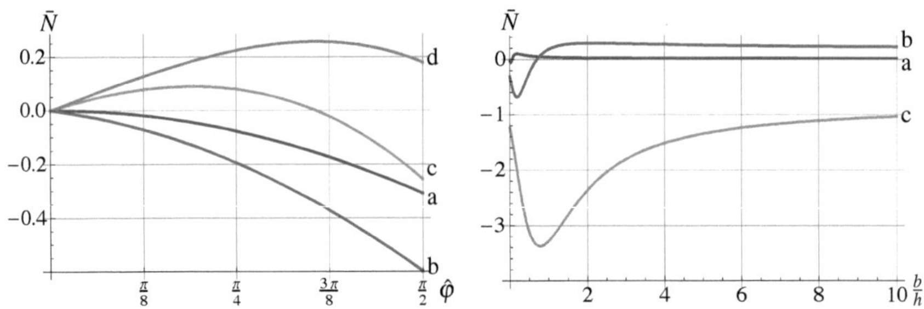

Fig. 4 Cylinder. Axial force VS $\hat\varphi$ for different pitch/height ratios: b/h=0 (a), 0.1 (b), 0.5 (c), 1 (d) (left). Axial force VS pitch/height at $\hat\varphi = \pi/2$ for different aspect ratios of the cylinder: $R_o/h = 0.1$ (a), 0.5 (b), 1 (c) (right). $\gamma = 1$ in both plots

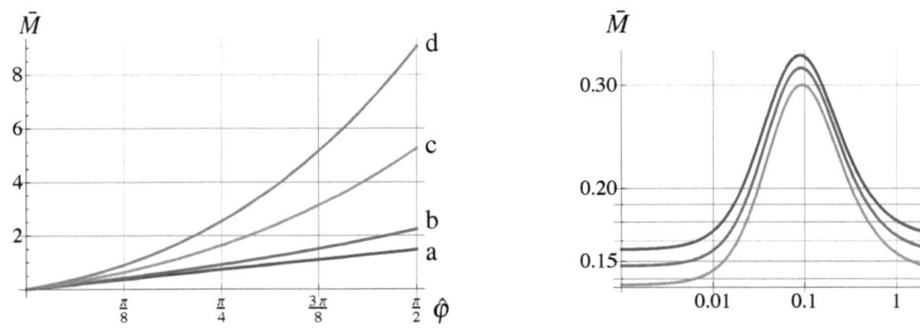

Fig. 5 Pipe. Torque VS $\hat\varphi$ for different fiber stiffness $\gamma = 0$ (a), 0.1 (b), 0.5 (c), 0.6 (d). Right: Torque VS pitch/height for different wall thickness of the tube: Ri/Ro=0.1 (a), 0.5 (b), 0.6 (c); Ro/h=1, $\gamma = 1$, $\varphi_h = \pi/2$ (right)

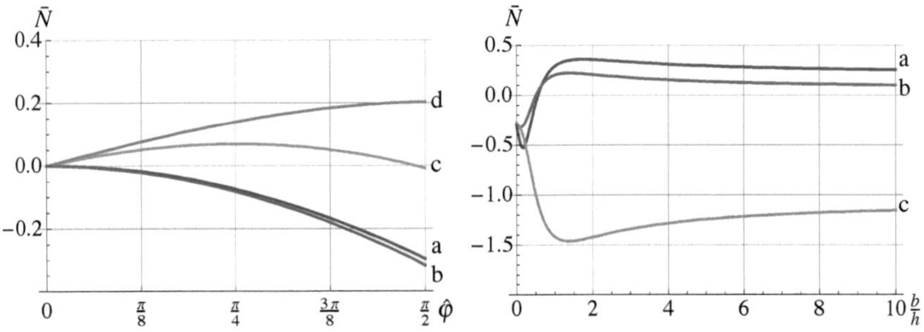

Fig. 6 Pipe. Axial force VS $\hat\varphi$ for different pitch/height ratios: b/h=0 (a), 0.1 (b), 0.5 (c), 1 (d), with $R_o/h = 0.5$, $R_i/R_o = 0.2$ (left). Torque VS b/h for different wall thickness of the tube: $R_i/R_o = 0.1$ (a), 0.2 (b), 0.3 (c), with $R_o/h = 0.5$, $\gamma = 1$, $\varphi_h = \pi/2$ (right). $\gamma = 1$ in both plots

and $\bar{N}$ on b/h for different slenderness parameters, at $\hat{\varphi} = \pi/2$ (right). The same analysis is performed for a pipe; in such a case, while $\bar{M}(\hat{\varphi})$ is always an increasing monotone function with respect to $\hat{\varphi}$ and γ, Fig. (5, left), more interesting in a pipe is the behaviour of $\bar{M}$ with respect to b/h at a fixed $\hat{\varphi}$ and γ ((5), right); as figure (5) evidences, it exists a specific and small range of b/h where $\bar{M}$ attains large values, actually independent on the ratio R_i/R_o. Dependence of the dimensionless axial force $\bar{N}$ on $\hat{\varphi}$ or on b/h is very similar for cylinders and pipes, cfr. Fig. (4) and (6).

References

1. de Botton, G., Hariton, I., Socolsky, E.A.: Neo–Hookean fiber–reinforced composites in finite elasticity. J. Mech. Phys. Solids 54, 533–559 (2006)
2. Merodio, J., Saccomandi, G., Sgura, I.: The rectilinear shear of fiber–reinforced incompressible non–linearly elastic solids. Int. J. Non-Lin. Mech. 42, 342–354 (2007)
3. Merodio, J., Ogden, R.W.: Mechanical response of fiber-reinforced incompressible non-linearly elastic solids. International Journal of Non-Linear Mechanics 40, 213–227 (2005)
4. Merodio, J., Ogden, R.W.: Tensile instabilities and ellipticity in fiber-reinforced compressible non-linearly elastic solids. International Journal of Engineering Science 43, 697–706 (2005)
5. Merodio, J., Ogden, R.W.: On tensile instabilities and ellipticity loss in fiber-reinforced incompressible non-linearly elastic solids. Mechanics Research Communications 32, 290–299 (2005)
6. Ogden, R.W.: Nonlinear elasticity, anisotropy, material stability and residual stresses in soft tissue. In: Holzapfel, G.A., Ogden, R.W. (eds.) CISM Courses and Lectures Series, vol. 441, pp. 65–108. Springer, Wien (2003)
7. Rivlin, R.S.: A note on the torsion of an incompressible highly-elastic cylinder. Proc. Cambridge Philos. Soc. 45, 485–487 (1949)
8. Spencer, A.J.M.: Continuum Theory of the Mechanics of Fibre-Reinforced Composites. Springer, Heidelberg (1984)

Author Index

Badel, Pierre   71
Boisse, Philippe   71
Braun, Manfred   1

Carillo, Sandra   63

Ganghoffer, Jean-François   49

Ilison, Lauri   21

Maire, Eric   71
Mangipudi, K.R.   29

Nardinocchi, Paola   37, 79

Onck, P.R.   29

Pastrone, Franco   9
Podio-Guidugli, Paolo   37

Salupere, Andrus   21
Svaton, Tomas   79

Teresi, Luciano   79

Vidal-Sallé, Emmanuelle   71

Program

Monday, March 10, 2008

08.00 − 9.00	Registration
09.00 − 9.45	Welcome

SESSION 1 - Chairman: Samuel Forest

09.45 − 10.30   **Paolo Podio Guidugli**, P. Nardinocchi
AFC: active tension vs active deformation.

10.30 − 11.00   Coffee break

SESSION 2 - Chairman: Luigi Preziosi

11.00 − 11.30   **P. Badel**, E. Maire, E. Vidal-Sallé, P. Boisse
Mesoscopic mechanical analyses of textile composites. Validation with x-ray tomografy.

11.30 − 12.00   **C. Barbier**, R. Dendievel, D. Rodney
Computational approach to the mechanics of entangled materials.

12.00 − 12.30   **P. Boisse**, Y. Aimène, P. Badel, S. Gauthier, E. Vidal-Sallé
Computational analyses of fibre composite reinforcement deformation using hypo and hyperelastic constitutive equations.

12.30 − 14.30   Lunch

SESSION 3 - Chairman: Jean Francois Ganghoffer

14.30 − 15.15   **R. Ogden**
Non-uniqueness and non-smooth solutions in the nonlinear elasticity of fibre-reinforced solids.

15.15 − 15.45   **D. Durville**
A finite element approach of the behaviour of entangled and woven materials at microscopic scale.

15.45 − 16.15   **P. Dyatlova**
Nonlinear problems of fibre reinforced soft shells.

16.15 − 16.45   Coffee break

SESSION 4 - Chairman: Alain Molinari

16.45 − 17.15   **P. Nardinocchi**, T. Svaton, **L. Teresi**
On the mechanical response of fiber-reinforced incompressible non-linearly elastic pipes.

17.15 − 17.45   **F. Moravec**, M. Holecek
Modeling the smooth muscle tissue as a dissipative micro-structured material.

17.45 − 18.15   **A. DiCarlo**, S. Naili, V. Sansolone
Growing poroelastic media.

Tuesday, March 11, 2008

SESSION 5 - Chairman: Paolo Podio Guidugli

09.15 – 10.00 **S. Forest**
The micromorphic approach for plasticity and damage in microstructured materials.

10.00 – 10.30 **A. Berezovski, J. Engelbrecht, G.A. Maugin**
One-dimensional microstructure dynamics.

10.30 – 11.00 Coffee break

SESSION 6 - Chairman: Philippe Boisse

11.00 – 11.30 **V. Libertiaux, F. Pascon**
Viscoelastic modeling of brain tissue: a fractional calculus-based approach.

11.30 – 12.00 **G. Del Piero, G. Pampolini**
Strain localization in polyurethane foams. Experiments and theoretical model.

12.00 – 12.30 **K.R. Mangipudi, P.R. Onck**
Fracture in open-cell metal foams: a multi-scale approach.

12.30 – 14.30 Lunch

SESSION 7 - Chairman: Arkady Berezovski

14.30 – 15.15 **G.A. Maugin**
On inhomogeneity, growth, ageing and the dynamics of materials.

15.15 – 15.45 **M. Istkov, A.E. Ehret**
An anisotropic softening model with application to soft biological tissues.

15.45 – 16.15 **J.F. Ganghoffer, N. Mefti**
A 3D stochastic model of the cell-wall interface during the rolling.

16.15 – 16.45 Coffee break

SESSION 8 - Chairman: Ray Ogden

16.40 – 17.15 **R.A. Abirov**
Investigation of strength and bearing capability of polymer pipes under complex loading.

17.15 – 17.45 **A. Burteau, F. Nguyen, J.J. Bartout, S. Forest, S. Saberi**
Investigation of the representative volume element size and analysis of the deformation mechanisms of open-cell foams.

17.45 – 18.15 **E. Caroli, G. Pallotti, P. Pettazzoni**
The mechanical structures of great arteries of cardiovascular system.

Wednesday, March 12, 2008

SESSION 9 - Chairman: Antonio DiCarlo

09.15 – 10.00   **F.J. Barthold, M. Rotthaus**
Analysis and optimization of interfaces for multimaterial structures.

10.00 – 10.30   **V. Marcadon, E. Roques, F. Feyel**
Modelling of mechanical behaviour of metal hollow sphere regular packings under compression loadings.

10.30 – 11.00   Coffee break

SESSION 10 - Chairman: Alexey Porubov

11.00 – 11.30   **D. Ambrosi**
Stress and growth in soft biological tissue.

11.30 – 12.00   **N. Kizilova**
A mixture model of the growing viscoelastic plant biocomposite.

12.00 – 12.30   **M. Kroon, G.A. Holzapfel**
An inverse method to estimate the material parameters and wall stresses of a saccular cerebral aneurysm.

12.30 – 14.30   Lunch

SESSION 11 - Chairman: Juri Engelbrecht

14.30 – 15.15   **A. Porubov**
Essentially nonlinear strain waves in solids with complex internal structure.

15.15 – 15.45   **L. Orgéas, J.P. Vassal, C. Geindreau, D. Favier**
Analysis of blood flow through tumoral microcirculatory networks via homogenisation.

15.45 – 16.15   **P. Cermelli**
Constant angle surfaces in nematic fluids.

16.15 – 16.45   Coffee break

16.45 – 18.15   Poster session

Thursday, March 13, 2008

SESSION 12 - Chairman: Gerard A. Maugin

09.15 – 10.00   **M. Braun**
Polar elasticity.

10.00 – 10.30   **N. Auffray, R. Bouchet, Y. Bréchet**
Derivation of anisotropic matrix for bi-dimensional strain gradient elasticity behaviour.

10.30 – 11.00   Coffee break

SESSION 13 - Chairman: Franz-Joseph Barthold

11.00 – 11.30   **Kh. Fataoui, B. Pateyron, M. El Ganaoui, P. Fauchais**
Simplified 2D transient modeling of splat formation: application to zirconia.

11.30 – 12.00   **C. Ittu, N. Constantin, C. Rizescu, F. Miculescu, P. Balog**
The analysis of slag used in casting steel process.

12.30 – 14.30   Lunch

14.30 – 20.00   Excursion

20.00–   Social dinner

Friday, March 14, 2008

SESSION 14 - Chairman: Franco Pastrone

09.15 – 10.00   **J. Engelbrecht, M. Randruut, A. Salupere**
On modelling wave motion in microstructured solids.

10.00 – 10.30   **S. Carillo**
Nonlinear Waves and Linear Heat Conduction with Memory.

10.30 – 11.00   Coffee break

SESSION 15 - Chairman: Manfred Braun

11.00 – 11.30   **V.I. Erofeev, A.V. Sharabanova**
Nonlinear Effects During Shear Waves Propagation in Polymeric Materials.

11.30 – 12.00   **P. Giovine, V. Giacobbe**
Plane Waves in Linear Thermoelastic Porous Solids.

12.00 – 12.30   **A. Salupere, L. Ilison**
Numerical simulation of interaction of solitons and solitary waves in granular materials.

12.30 – 13.00   Farewell

MIX
Papier aus verantwortungsvollen Quellen
Paper from responsible sources
FSC® C105338

If you have any concerns about our products,
you can contact us on
ProductSafety@springernature.com

In case Publisher is established outside the EU,
the EU authorized representative is:
Springer Nature Customer Service Center GmbH
Europaplatz 3, 69115 Heidelberg, Germany

Printed by Libri Plureos GmbH
in Hamburg, Germany